Contents

I	Meat Preserving, Past and Present	5
II	Salted or Pickled Meats	9
III	Smoked Meats	18
IV	Frozen Meats	21
V	Meats Sealed in Fat	29
VI	Spiced Meats	33
VII	Brawns and Head Cheese	40
VIII	Sausages	46
IX	Potted Meats	59
X	Pâtés and Terrines	64
	Useful Reading	72
	Consumer Guidance Organisations	72

Meat Preserving Past and Present

Man almost certainly learned to preserve meat before he learned to cook it. Probably he learned by accident to dry it, and maybe to smoke and salt it too, in very ancient times when he was still a hunter. The advantages quickly became clear—he could keep the meat he could not eat straight away, and he must soon have made a habit of hanging up the surplus meat to dry, out of the reach of other carnivores.

He must have discovered, also by trial and error, that ashes from the fire (which contained potash) would also preserve meat. In places where salt was plentiful carcasses found intact in the sea or salt-beds indicated the preserving power of salt, soon to become one of man's most precious commodities, while meat left hanging in the smoke from his fire also had a longer life and better flavour.

Most of us probably think of deep freezing as a slick modern development. But to the peoples of the distant north, freezing was the obvious way to store meat, for they saw it being done naturally all around them. A bear trapped in the ice, a fish left in the snow, could be picked up and eaten months later.

Drying, salting and spicing, smoking, cooking, freezing! All these ways of preserving meat were known far back in history, and have been in use ever since. Only one really new method of meat preserving has been added in modern times. That is bottling and canning. The basic idea of preserving food in containers was first developed by a French scientist about 1805 because Napoleon wanted to find a way of feeding his troops on the march without relying on local foods.

Having to rely on local foods was the reason why meat preserving was so vital. Man would have starved without these skills. So whenever a beast was slaughtered for food, every part of it not to be eaten at once was processed, either by the butcher or in the homestead kitchen. In any household, rich or poor, sides of pork were cured for bacon. Legs were cured too, for hams, and were hung in the chimney to smoke. Other meats and poultry were also dried by smoking, or were pickled in salt, as tongues and silverside still are today.

Eating dried and salted meat for months on end was monotonous, so to make their preserved meats more interesting, people developed different ways of using them. As far back as 8000BC, the Greek poet Homer mentioned sausages. By Roman times, sausages were so popular that various kinds were made, in great quantities; in fact, our own word 'sausage' comes from the Latin *salsus* meaning 'meat preserved in salt'.

In cities, from Roman times onwards, specialist pork butchers made most of the sausages, and places like Frankfurt in Germany, Bologna in Italy, and Lyons in France, created their own variations and gave their names to them. But in every smaller town and village, housewives took pride in making their own sausages too, and developed other means of using preserved meats as well. For instance in Britain, the art of collaring fresh meat was devised, a means of short-term storage and use found nowhere else. Sweet cured meats, and mixtures of dried fruit and pounded meat (which later became our mincemeat) were popular too. On the Continent, in France especially, housewives preferred to seal cooked meats in fat; and from this, French housewives came to make the delicious *confits*, potted meats and pâtés which are still such an important part of their cookery, and of our own. There are sections on them later in this book, as well as one on sausages.

The old housewives knew ways of preserving every part of a carcase, even offals, fat and blood. As short-term preserves, brawns and 'head cheeses' were made, fat was rendered, and the crisp fatty scraps which have various names in different parts of the country, were kept and used. Blood puddings were made too, our own 'black puddings' among them.

Using every part of a carcase was commonsense economy, even in large wealthy households, until 100 years ago. With today's rising meat prices, it is, again, something we should all be able to do. Over the centuries, housewives of the past acquired great skill in all the ways of preserving meat at home, both long-term and short-term. They knew from experience how to dry, cure and smoke meats successfully as well as make the short-term preserves which made their meals interesting at little cost. We must learn from them if we can.

World shortages of certain foods combined with spiralling processing costs make any practical modern cook look for new ways to maintain the standard and variety of his or her cuisine. Although some of the skills of former days are no longer safe or practical to revive, many others are. So this book looks at the old crafts, and tells you which ones are practical and useful. Then it describes the ways of preserving meats which you are likely to find useful and gives you recipes which are safe and sensible for them.

Large-scale and long-term meat preserving is not a practical proposition for most of us. For one thing, even if we had the tools, space and skill to do it, we could not afford and would not need to cure whole sides of bacon or several hams. Getting the tools alone makes large-scale preserving too expensive, but besides this, long-term preserving of modern meats is difficult without a great deal of experience, and is likely to be dangerous to do at home. We know much more these days about the harm bacteria can do, and because of this, stringent government regulations control certain processes.

Short-term preserving, however, of smaller items, such as 'special offers' or unused meats or offals, is practical and can certainly help to stretch our budgets. Today, we may not starve without the pigs' ears, brains and tongues, the oxtails and calves' feet, breast of lamb and sheeps' hearts; but some of them if properly preserved and used, make economical and delicious meals.

Finally, as well as bringing you good

eating and cost consciousness, the revival of these old skills brings other advantages too. They give you the opportunity of using home produced raw materials to full advantage, and also allow you to control the amount of added artificial bulk foods you eat as well as your intake of the artificial preservatives, colouring and so on found in so much pre-packaged food today. So home meat preserving also helps you to make a positive contribution to healthier eating—providing that you observe all the safety precautions as well, of course! Not least, too, the revival and practise of these old crafts can be a relaxing and amusing hobby.

Note on Measures

The quantities in the recipes are given in metric measures with British Imperial measures in brackets. The spoons used are British Standard measuring spoons. They should always be measured level, *i.e.* with the contents smoothed off level with the rim.

Standard Cuts of Beef

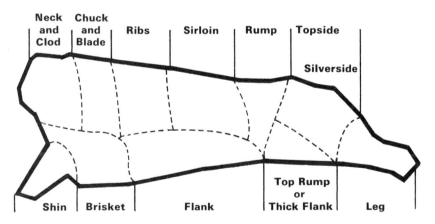

The better cuts

Sirloin
Rump steak
Ribs, including Fore ribs,
    Wing ribs, Back ribs
    and Top ribs
Topside

The economy cuts

Top rump (also called
Thick flank, Bed of beef,
First cutting)
Brisket
Silverside
Shin

Leg
Neck and clod
Flank
Chuck and blade
Skirt

British Meat Service

Standard Cuts of Lamb and Pork

Standard Cuts of Lamb
British Meat Service

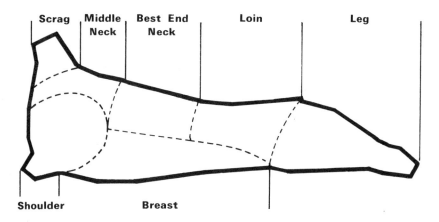

The better cuts

Saddle
Leg, including Fillet-end, and Shank- or Knuckle-end
Loin and loin chops
Chump chops

The economy cuts

Shoulder, including Blade End and Knuckle End
Best end neck
Middle neck
Breast
Scrag

Standard Cuts of Pork
British Meat Service

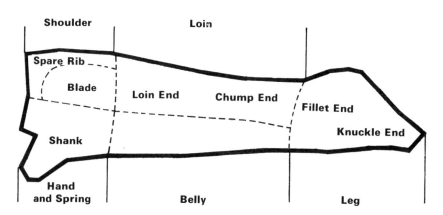

The better cuts

Leg
Loin, including Loin chops and Chump chops
Tenderloin (also called pork fillet)

The economy cuts

Shoulder, including Blade and Spare rib
Hand and spring, including Hand and Shank
Belly (also called draft or flank)
Head, and trotters

Salted or Pickled Meats

If salt really penetrates into a foodstuff, it prevents destructive bacteria growing in a way that nothing else does. So we can preserve foods in salt alone, and whatever else we use, we cannot pickle meat without it.

'To pickle' only means 'to preserve'. So pickling really just means preserving in salt. The other substances we add to pickles certainly help to preserve foods, and also give them a more interesting flavour. But, by themselves, they cannot keep food free of decay and fit to eat.

The most common pickling agents besides salt are vinegar, pepper, alcohol in the form of wine, spirits or beer, and wood smoke which contains alcohol and tar among other things. Some pickling recipes use them all, and various flavouring and colouring spices as well.

Salting or pickling, together with drying, are what we call 'curing'. It is a very old form of preserving indeed. The Roman writer Cato described it in 200BC, and the method of doing it has not changed since his day. Salting or curing meat has always been done at home, too, so there is no reason why we should not do it now. The only real difficulty is that we have forgotten how to do it because our smaller households do not need whole sides of bacon, a row of hams hanging in the chimney, and various other salted meats besides. Nor can most of us afford the joints!

There are, however, ways in which home-salted and pickled meat can be useful nowadays, provided you are scrupulously careful about how you do it, to prevent the meat being tainted.

One way in which home-salted or cured meat is useful is that it lets you take advantage of cheap meat by buying in bulk ahead of needing it, like meat for freezing (page 21). For people without a freezer, pickling can be a boon. It is also helpful for people who live in remote areas, or who are always busy and only want to shop once a month.

Another attraction is that many more kinds of meat can be pickled than we usually imagine, and items like home-pickled beef, lamb, duck or tongue can give a welcome variety to meals. Besides, even if you can buy a pickled tongue from a butcher or supermarket the flavour of the home-produced meat will be different and usually more intriguing; and by changing the pickle recipe slightly, you can change the flavour as you wish.

There is another advantage of home curing. Largely because factory-cured bacon and ham in particular are processed and sold for quick use, they are only very lightly cured today. They are also lightly cured because we prefer leaner, milder-flavoured pork meats than people did in the past; so comparatively little salt is used. The disadvantage of this light modern curing is that factory-cured bacon, gammon or ham which is not used soon becomes unattractive to use. It may smell rancid and be dry and tough to cook. In a family whose movements are uncertain, it can be useful therefore to have a few pieces of more heavily-cured meat in stock, which can be taken out of the pickle and used as and when they are wanted.

Joints of pickled meat must be soaked before use as a rule (depending on the recipe), but this takes no longer than thawing most frozen meats. So, if you have suitable conditions for pickling and storing the products, these meats can be an excellent household standby.

Both the right ingredients and the right conditions are essential for successful home pickling. Take great care to make sure that they *are* right because being careless can create dangerous hazards.

Ingredients for curing

Meat

The first and most important ingredient is of course the meat. Whatever meat you use, remember that in these days there is no need to pickle large joints. Do not let old recipes with their descriptions of whole hams and large quantities of pickling ingredients frighten you. It is just as practical to pickle one or two 1 kg (2 lb.) pieces of pork belly as a large 5–6 kg (12 lb.) gammon. They will give you two or three good family meals and breakfast slices to fry; and then you can use a slightly different pickle for the next cure.

Other advantages of pickling small pieces are that you can use a lighter pickle, with less risk of making the meat oversalted. It will take less time to cure, and you can use domestic tools and containers instead of buying special equipment and tubs as you would have to do for large joints.

You cannot buy or cure less than a whole ox tongue. But a maximum weight of $1\frac{3}{4}$–$2\frac{1}{4}$ kg (4–5 lb.) per piece of meat should be your aim, at least until you are experienced in the craft of curing.

The weight of the meat matters less, in fact, than its shape and thickness. Solid thick pieces of meat take a good deal longer to pickle than thin ones because the salt takes longer to penetrate to the centre of the meat. So it is important, if you are pickling parts of a home-killed or farm-bought whole carcase or several pieces of butcher's meat at once, to take the thinner pieces out of the pickle first. It will probably be convenient to do so in any case, since you will not want to use all the meat at once.

Do not buy very lean meat. Since fat absorbs less salt than lean meat, fatty meat remains softer and tastes less salty after curing. Wherever you get your meat from, too, pickle it as soon as possible after it is cut into pieces. This will give bacteria less chance to attack the cut surfaces.

If you buy abattoir or butcher's meat, you need not worry, today, that errors in slaughtering or handling may cause faulty curing. But if you get meat from a farmer or other private source, try to check that the animal has been handled correctly before and after slaughter. Bulletin 127 of the Ministry of Agriculture and Fisheries, titled *Home*

Curing of Bacon and Hams contains valuable information about this. Unfortunately, it was out of print at the time this book was written, but a good public library should be able to supply a copy.

Salt

There are four kinds of salt which can be used for curing. All have some advantages and disadvantages, but any of them will cure the meat successfully, and they should be obtainable from any good store.

Bay salt is made by evaporating brine in large, shallow, open pans. It is a coarse, hard, damp salt with large crystals. It gives a wetter cure than the other salts because its big crystals do not mop up as much surface liquid on the meat.

Common salt is made in the same way as bay salt but under different evaporating conditions. It has finer crystals, is light in weight and easy to handle.

Bar salt is heated, packed into moulds and then dried. Dairy salt is the same salt crushed and sieved. Both are heavier than common salt.

Vacuum salt is made in a different way from other salts, in closed containers without air. It is purer than other salts and has fine, small, even crystals. It may seem the best to use because it is purer, but it is heavier to handle than the other salts, and its fine crystals tend to cake on the meat's surface. When the meat shrinks it may no longer be touching the salt covering, and bacteria can grow in the space between them.

To prevent this happening, many people like to use a mixture of coarse and fine salt.

Whichever salt or mixture of salts you choose, remember that even though you are processing a small piece of meat, it will need more salt, and will taste more salty, than factory-cured meat. Recipes for home-cured meats contain the extra salt because, as a rule, one uses the meat less quickly than a small quantity of butcher's meat bought for one or two immediate meals. The longer meat is to be kept, the more heavily cured it must be.

You can use salt or a salt mixture on meat in one of four different ways:

1. Soak the meat in heavily-salted liquid or brine, using a 'wet cure'.
2. Rub the meat thoroughly with a lot of salt. This will draw quite a lot of liquid from the meat. By letting this liquid lie on the surface of the meat or by basting the meat with it, you get almost the same kind of 'wet cure' as by using brine.
3. Sprinkle the meat with a little salt fairly often, and only rub it in lightly. The meat does not give out much liquid, but its surface is always moist.
4. Pack the meat down into a bed of dry salt and cover it completely with more salt. All the liquid drawn out of the meat is absorbed by the salt so that the meat itself is always dry.

All these ways of salting will cure the meat, but they will not do it in quite the same time or the same way. The one you choose will depend on the type of meat, on when you want to use it, on where you can cure and store it and on the flavour you want it to have.

Wet curing is quicker than dry curing; the salt penetrates the meat more quickly. The meat is less hard and salty too, which helps if you have not much time to soak it before cooking. On the other hand, dry curing is more certain and preserves the meat longer.

Wet curing gives you a wider choice of flavours. You can rub dried herbs into the meat with dry salt, and saltpetre if you use it, but if you use sugar or treacle, wine or beer, to flavour your meat, it will make a wet cure.

Among other pickling materials, the one you see mentioned most often in recipes is saltpetre, or the form of it called 'salprunella'. Saltpetre turns the meat a rosy pink colour instead of dingy grey as salt does alone. But as saltpetre can make the meat very hard, sugar is often added to keep it soft. Sugar seems to help prevent the meat decaying too.

Very little saltpetre is needed to give meat a pink colour, so do not use more than the recipe states; you will only risk making the meat hard.

Another way to colour the meat is to use a treacle cure. It gives the meat a dark, rich colour which is especially attractive on fat pork and in sausages.

Tools you will need

You do not need elaborate equipment for curing. It is possible to manage quite well just with ordinary kitchen tools and pans if you only pickle small pieces of meat. Here is a list of what you will need:

1. Tenon saw.
2. Kitchen knife with a narrow point, or a boning knife.
3. Carving knife or large steak knife.
4. Scales for weighing meat and other ingredients.
5. Strong kitchen table.
6. Large basin or tub for wet curing (or a trough with a plug-hole in the bottom).
7. Large low-sided box or tray for dry curing.
8. Stone or slate shelf or floor for dry curing.
9. Sterilised (boiled) board and weights to hold down the meat (large stones make good weights).
10. Meat hooks or very strong wire hooks.
11. Strong twine or string.
12. Strong cotton material for storage bags (an old pillowcase makes a good bag for a large piece of meat).

If you buy ready-cut meat from a butcher, you may not need a saw for cutting through bone (1), although it is useful for trimming jagged bone ends. The boning and steak knives mentioned (2) (3) are butcher's tools, but good quality, sharp kitchen knives do quite well for cutting up or trimming small joints and other pieces. A trough with a plug-hole (6) will be useful if you want to try method two of applying salt (page 11 above). You can drain off, through the plug-hole, the liquid drawn out of the meat by the salt, and use it for basting. If you prop up one end of your tub or trough so that the bottom is sloping, you can put thicker pieces of meat in the lower end where the pickle collects.

Where to Cure the Meat

Although you do not need a special room for pickling the meat, it is very important to have the right conditions, to be sure that your curing is safe and successful.

Heat makes salt penetrate the meat faster, but it makes bacteria multiply faster too. At very low temperatures, salt may take a long time to penetrate to the bone in a thick piece of meat, giving bacteria a chance to develop in the centre of the meat. So try to cure meat at a temperature between 2°C and 10°C (35°F and 50°F), if possible in a north-facing room or larder. A cellar, although cool enough, is usually too damp except for dry curing in a bed of salt (method four on page 11). Keep a thermometer in the room to check the temperature. The room you use should have plenty of fresh air, but you must screen the windows against flies. Obviously, you need to be able to keep out dogs or cats, rats and mice too. You will also need curtains or blinds to keep the room fairly dark, since it helps to prevent the fat becoming rancid.

Ideally, the room should have a stone or cement floor, or a slate shelf on which the curing containers can stand.

For drying the meat after pickling it, you need a room with an even temperature of about 16°C (60°F). Any passage or room where you can hang up your meat hooks will do, provided it gets some fresh air and is dry.

If you want to store the meat, instead of using it straight from the pickle or as soon as it is dried, keep it in a cool, dark, dry place with an even temperature. Do not put it in an attic or cellar. The temperature in an attic usually varies a good deal, and a cellar may be damp with too little air. Under any of these conditions, the meat may get damp and even slimy. Rather keep it in a north-facing larder or some other unheated room.

How to Cure Meat

If the meat has not been cut up ready for pickling, lay it on the table (5 in the list above) and use the saw (1) and knives (2 and 3) to prepare it. In any case, trim it neatly, leaving no ragged ends of flesh or skin. Then weigh the meat and record the weight, especially if you want to use the General Use Dry Salt Cure. The meat will lose weight in curing, but it will only lose moisture, not food value.

Now calculate how long you will leave the meat in pickle, that is in salt, salt and saltpetre, or salt and other ingredients. To do this, measure how thick the meat is. Most moderately fatty meat needs about seven days in pickle for each 2.5 cm (1 in.) of its thickness to be fully pickled. Properly dried and stored, it will then keep for six months. (Note, though, that modern tastes usually prefer a less 'hard' cure than this, although the meat does not keep as long.)

Record the date in a book or on a wall chart, to remind you when to take the meat out of the pickle. You can, of course, take it out sooner, provided you use it within a short time. You can also leave it in the pickle longer if you do not want to eat it, or to dry and store it, as soon as it is ready. The longer you leave it in the pickle, the longer it will keep afterwards. But it will go on getting harder and saltier and may need long soaking when you want to use it. In any case, you should not leave it in pickle for longer than five weeks.

Next, read the recipes below carefully. Decide which you will use. It will depend on the weight, thickness and fattiness of the meat, the materials you have for curing, when you would like to use the meat and the flavour you want to give it. Check that you have all the materials and tools you need and that your containers are clean and dry. Check the temperature of the room.

Pickled gammon — British Bacon Curers Federation

Now follow your chosen recipe carefully. All the recipes are tested and safe for anyone to use, given the right materials and conditions and provided the instructions are carried out in detail.

General Use Dry Salt Cure

Pieces of meat weighing not more than 3 kg (6½ lb.) each and of the same thickness
Salt for pickling tray (see recipe)
Strong brine for cleaning meat (see below)
One-tenth of the weight of the meat in mixed coarse (bay) and fine (common or vacuum) salt
One-fortieth of the weight of salt in saltpetre

The usual proportions for making the brine are:
10 litres (2¼ gallons) boiling water
3 kg (6½ lb.) salt
25 g (1 oz.) saltpetre

Dissolve the dry ingredients in the water. Let it cool to 10°C (50°F), then place the meat in the brine. Leave flat pieces of meat such as belly of pork in the brine for 10–15 minutes. Leave pieces with corners, such as a fore-hock of pork, for about 30 minutes. After soaking, remove the meat and lay it on paper on the table to drain. If you want to use a wet cure for some small pieces of meat while dry curing the bigger pieces, add $3\frac{1}{2}$ litres ($\frac{3}{4}$ gallon) water to the brine for a pickling medium.

Now prepare the dry salting mixture. Divide the dry mixed salt into three equal portions. Keep them separate; for rubbing in; for sprinkling; and for repacking. Crush and sieve the saltpetre and add one-third of it to each pile of salt.

Rub the meat with the first portion of salt and saltpetre. Rub it well in on the skin side, more lightly on the flesh side. If the meat has a bone, rub salt well in round it and up into the hollow of the bone.

Now prepare a bed of salt in a low-sided box or deep tray if possible big enough to take the pieces of meat in one layer. It must be at least 5 cm (2 in.) deep. Press the pieces of meat into it, skin side down. Sprinkle them with the second portion of salt and saltpetre. Then cover them with about 5 cm (2 in.) salt and pack it tightly, close to the meat, at the sides. Leave for five days.

After five days, break up the salt covering the meat and remove any discoloured salt. Repack the meat, with the pieces you will use first where you can get at them easily. Before covering the meat with fresh dry salt, sprinkle it with the last portion of salt and saltpetre. Then cover the pieces, packing salt tightly around the sides of the meat as before. Leave in the salt until the end of the calculated time, or until you need them (see above). Each time you remove a piece of meat from the salt repack the rest tightly with their 'jackets' of salt to make sure that the salt is in contact with the meat all the time.

You can take out a single piece of meat when you need it, or remove all the pieces when ready and dry them.

You will probably have to soak any piece of meat out of this dry salt pickle before cooking and eating it, whether you dry and store it first or not. How long you soak it will depend on the thickness of the meat, how fat it is and how long it has been in the salt. Times can vary from an hour or two to two days for a thick, heavily cured piece. Soak the meat in clean fresh water, which should be changed when it gets very salt. Then weigh the meat and boil it in the same way as any pickled meat (*e.g.* pickled pork, ox-tongue or silverside) bought from a butcher. Any big standard cook-book supplies recipes.

To Dry the Meat
If you take all the meat out of the salt bed at once, you will have to dry and store any pieces you do not want to use at once. First, wash them in cold water to remove the salt clinging to them. Dry them thoroughly. Then hang them on meat hooks in a temperature of about 16°C (60°F) as suggested on page 13. As the meat dries, a white layer of salt may appear, especially if the room is not as dry as it should be. Wipe this off carefully as soon as it appears.

To Store the Meat
When the meat is really dry, you can store it. It is best stored in the dark, at a temperature not above 16°C (60°F), but the most important thing is to store it at an even temperature.

If the meat is well dried and can be

kept dry, you can get away with not covering it. But it is best protected by being hung in a calico or linen bag, tied up tightly with twine, or by being buried in clean oatmeal. The length of time it will keep will depend on how long you have kept it in pickle and how well dried it is. But do not keep it longer than you need.

If flies or other pests attack meat hanging up while it is stored, dust it with black pepper, especially round corners and bone ends. If the meat is stored in oatmeal, make sure the container can be closed firmly to keep out mice.

The meat should not become slimy or grow any mould while being stored. Slime means that it is damp. Dry it really well and place it in a dryer atmosphere. Mould on the meat may mean it has gone bad and it is usually best thrown away.

In just one case, however, you can eat such meat safely. If you cut a piece of cured meat and put the rest back into store, the cut surface may grow a white mould with a strong ammonia smell. This is very like a Camembert cheese mould. Cut off the mouldy layer of meat and throw it out. You can eat the rest safely.

Wet Pickle for Beef, Pork or Ox-Tongue

$1\frac{1}{4}$–$1\frac{3}{4}$ kg (3–4 lb.) meat
Common salt for rubbing
$2\frac{1}{2}$ litres (2 quarts) water
75 g (3 oz.) coarse raw salt
$\frac{1}{2}$ × 2.5 ml spoon ($\frac{1}{4}$ oz.) saltpetre
 (1 × 2.5 ml spoon or $\frac{1}{2}$ oz. for tongue)
150 g (6 oz.) bay salt (350 g or 14 oz. for tongue)

Rub the prepared meat well with salt and leave aside for 24 hours. Boil all the other ingredients together until no more scum rises. Skim well. Allow the pickle to get quite cold. Place the meat in a clean crock or tub and cover it entirely with the pickle. Weight it down; for instance, place a board on it with a large clean stone on top. Unless the weather is very hot, you can leave the meat in the pickle for up to five weeks.

Then (or sooner if there is any sign of white mould on the pickle) reboil the pickle with 1 × 2.5 ml spoon ($\frac{1}{2}$ oz.) sugar and 50 g (2 oz.) common salt. Skim it as before, allow it to get cold and place the meat in it, well weighted down, to finish curing if it is not yet ready.

You can use the pickle for more than one batch of meat, provided you reboil it with the same amount of extra sugar and salt after another five weeks, but do not reboil it more than twice.

Treacle and Ale Pickle for Pork

2 × 1 kg (2 lb.) pieces belly of pork both the same thickness
Bay salt for rubbing
250 ml ($\frac{1}{2}$ pt.) ale
250 ml ($\frac{1}{2}$ pt.) stout
250 ml ($\frac{1}{2}$ pt.) black treacle
150 g (6 oz.) bar salt
150 g (6 oz.) coarse bay salt
$\frac{1}{2}$ × 2.5 ml spoon ($\frac{1}{4}$ oz.) saltpetre

Rub the meat well with coarse salt and lay it aside for 24 hours. Boil all the remaining ingredients together for five minutes. Put the meat in a crock or tub. Let the pickle get quite cold, then pour it over the meat. Keep the meat in the pickle for up to three weeks, turning and rubbing it very day. Drain it, dry it well as suggested on page 15 above and store it wrapped in thin cotton or muslin. Use it within two months.

This pork becomes a fine bronze colour and makes very good fried breakfast rashers.

Spiced Sweet Pickle for Ham

About 10 kg (20 lb.) pork, *e.g.* 2 hams
1 kg (2 lb.) crushed bay salt
500 g (1 lb.) common salt
100 g (4 oz.) ground black pepper
50 g (2 oz.) saltpetre
50 g (2 oz.) dry English mustard
$1\frac{1}{4}$ kg (3 lb.) black treacle or coarse raw brown sugar or a mixture

Trim the meat. Mix all the remaining ingredients except the treacle and/or sugar. Rub the meat thoroughly with the mixture and place it in a clean crock or tub. After two days it will have given off a good deal of liquid. Mix in the treacle and/or sugar and allow the meat to lie in the mixture for up to five weeks, turning and basting it often. A ham will need the full time in the pickle, smaller joints not so long.

Drain, dry and store as above.

Smoked Meats

Housewives in past days usually smoked their larger joints of meat, especially bacon and ham cuts. Sides of bacon and whole hams hung from hooks in the farm smoke-house or wide chimney, being preserved for eating months ahead.

Smoking dries meat more thoroughly than hanging it up to dry naturally (page 13 above). It also helps to preserve it because the smoke contains formaldehyde, alcohol, tar and other substances which are preserving agents. Some of them give the meat the typical 'smokey' flavour and aroma as well.

Large-scale smoking is not, however, practical for most ordinary households now. Few of us have a high, wide old farmhouse chimney in which to hang whole hams, or even the space to build a small, outdoor smoke-house. Nor have we the old traditional skill needed.

Smoking solid foods is not easy. The fire must smoulder without ceasing and without flaring up; it must be watched and attended to. Solid joints, such as hams, cannot just be left to hang in the smoke, either. They must be smoked for a short, carefully calculated time each day or every other day; and each time they are taken down from their hooks, they must be rubbed with pepper, spices and herbs such as crushed bay leaves while still warm so that the herbs cling to the softened fat. It is heavy, laborious work, going on over three weeks or more.

Skill is needed, too, in order to make sure that the temperature of the smoke does not rise above 32°C (90°F). If it does, the fat on the meat begins to melt and it is spoiled. Yet, at the same time, the meat must be smoked enough to prevent any risk of mouldiness and spoilage from inadequate drying.

Big joints should not be smoked with any home smoking gear, not even in the bigger type of smoke 'oven' commonly made from a 45 litre (10 gallon) drum. A permanent or semi-permanent smoke-house is needed, built of brick or metal, with grids and hooks to carry the meats. A means of creating and directing a draught is also needed, with controllable air outlets and baffles to disperse the smoke into every corner of the 'oven'.

These items can be quite complicated and costly to install and to keep working efficiently. Yet without them, the meat may well only get partly smoked.

Besides these safety requirements, government and EEC regulations must also be met. Some fuels are now thought to produce cancer, and the rules are designed as protection against such dangers.

Given all the risks, it is probably

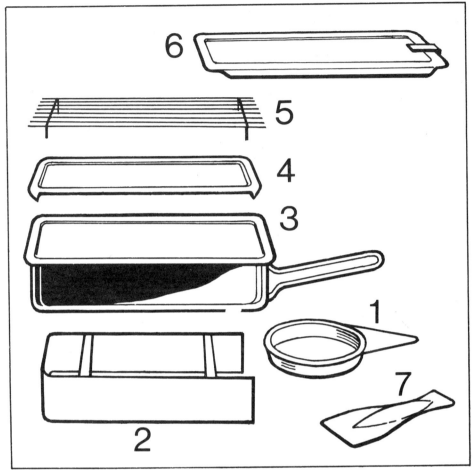

Diagram of Home Smoking Kit

1. Fuel-container
2. Wind-shield of aluminium
3. Smoke-box of whole-pressed aluminium
4. Dripping-cover of aluminium
5. Smoke-grate of stainless steel
6. Sliding lid of aluminium
7. Cleaning scraper—lid remover

Home smoking kit

Habitat Ltd and Henry Hulthen Ltd

wiser to take any solid meat joint or bird to a local bacon factory and arrange for it to be smoked there under controlled conditions. (Alternatively, instead of smoking, scrape and dry it well when it comes out of the pickle and rub it down with brandy before storing to preserve and flavour it.)

There is, however, one way in which home smoking can give you delicious meat dishes which are cheap and safe to make. You can smoke sausages and patties, home-made or bought, fresh or frozen, with excellent effect. They are moister than fried, grilled or baked ones, have a fine glossy chestnut colour while hot and make first-class cold or reheated meals or snacks.

Compared with buying Continental smoked sausages, our ordinary 'bangers' are cheap, yet they can be equally good when smoked in a small home smoking kit. There is a British make on the market which is particularly easy to use. A Swedish one is pictured. (See page 19.) A completely safe fuel is supplied for these 'kits' and the smoke level is well below the maximum strength allowed by EEC regulations.

The design of these home smokers is simple. It consists of a metal smoke-box or 'oven', a stand to put it on and a metal container for methylated spirit to put underneath. A drip-tray fits inside the smoke-box and a grid is supplied to stand on it, inside the box, to hold the food. Lastly, a sliding lid keeps the smoke within the box.

The system of using one of these home smoking 'kits' is equally simple. Scatter a layer of fuel in the bottom of the smoke-box, cover it with the drip-tray and a layer of foil, then with the grid. Prick sausages with a fork two or three times and season hamburgers or beefburgers with salt and pepper. Place them on the grid and cover the box with the lid. Fill the spirit container almost to the brim with methylated spirit, light it and push the container under the box.

Each container of methylated spirit gives about 15 minutes 'smoke' and one scattering of fuel gives 1–2 'smokes'. Full-sized sausages (8 in. 500 g or 1 lb.) may need turning after the first 15 minute 'smoke', being done after two 'smokes' or in about half an hour. There is no need to wait between the smoking periods. In fact, for a hot meal straight from the 'smoker' they are best fully smoked without a pause.

These sausages and patties are lightly cooked inside when done; first because the home smoking 'kit' gives a warm (technically a 'hot') smoke; second, because, since their texture is loose, the sausages absorb the heat. You can see this because some of their fat runs out, although less than when they are cooked by ordinary methods. This is fine if you want to eat them hot straight away, or fridge-chilled and reheated quickly, but if they are to be used for packed or picnic meals, they must first be reheated quickly to a high temperature and cooled quickly by being placed in a well-chilled container standing in cold water. In any case they should be frozen if not eaten straight away.

Any loose-textured mince, coarse pâté mixture or thin slice of meat can be smoked successfully in one of these smoking 'kits'. But do not try to smoke a single piece of meat such as a chicken joint. It requires longer, spaced-out smokings and the 'hot' smoke (as opposed to a 'cold' smoke of under 32°C (90°F)), will give the outside a pungent tarry taste before the food is smoked right through. In fact, 'hot' smoking may well form a dry coating on solid food so that later smoking does not penetrate it at all.

Thin slices of home-pickled pork get an interesting smokey flavour when processed in one of these 'kits'. Home-made sausages and ''burgers' can also be smoked successfully. Recipes for making them are given in chapter VI, page 46.

Frozen Meats

Modern freezing is the most convenient, quickest and safest method of preserving meat. Yet, in some ways, our techniques are much the same as the ones used by primitive man.

For instance, we can buy our meat in bulk or on a planned schedule, or just when it suits us. But so could he. He could let the beast he had hunted and killed remain frozen, whole, where it dropped, until he needed it; or before it stiffened he could cut off the pieces he wanted to eat at once and carry them home, coming back for the rest later. We have scientifically designed wrappings to save the meat from freezer burn, cross-flavouring and other defects. His carcasses had natural unbroken coverings of skin or fur. Granted we have a wider choice. He had to freeze and use any game which happened to cross his path, whereas we can choose from whatever the frozen food centre, multiple or private butcher or a local farmer has available; fat or lean, a lot or a little, as we wish. But, to balance that, we have to buy a large and costly home freezer to house our foods whereas in the Arctic at least he had all around him the immense one supplied free by Nature. Moreover he did not have to balance the cost of one meat cut against another or study how to use each piece of a carcase profitably; he used it all without thinking about it, because he had to.

Buying Meat

Wherever you propose to get your meat for the home freezer (assuming you will not butcher it yourself), you must know three things before you try to buy it: How much of each type of meat you will require; how you will use it; and the price you must expect to pay.

For instance, a whole or half carcase may look an attractive buy from some points of view; but much of it may be wasted if your household will not, under any circumstances, eat offal or if it includes a lot of unwanted fat or bone. You will waste meat too, if you buy, package and store larger quantities than you usually need.

As for calculating the right price to pay, there are three main things to consider.

1. The way a carcase is cut up varies a lot in different places. The meat will be more expensive than it need be if you order cuts which are not much used in the area where you live. Even if you find a butcher who knows the cuts and can let you have them, he will waste meat because they will not match what other customers want, and he will charge for this as well as for his specialist service.

2. The meat will also be more expensive if you order meats or cuts which other customers either like very much or do not buy at all. If lamb is very popular it

will be a lot more expensive than pork. If roasting joints are popular but liver and brains are not, then the shop may not stock liver and brains at all; they will have to be specially ordered if you want them. If fat meat is preferred to lean by a particular butcher's clients, this too will affect the price he charges, compared with a butcher in, say, the next street. So it is always worth window-shopping for the kind of meat you want.

3. Prices of meat vary seasonally. One of the advantages of preserving meat, especially by freezing it, is that you can buy it at its cheapest, to use at a time when it costs more. Most people like warming stews in winter rather than summer, so the price of stewing cuts rises in winter. The freezer owner will therefore gain by buying stewing cuts and oxtails at the warmest time of year. The same is true of pork which many people find too rich in spring and summer compared with, say, lamb. The quality of lamb is best in spring and summer but it is a good deal cheaper in autumn. Other hints on pricing meat are given under Bulk Buying below.

Bulk Buying

If you have a large freezer, bulk buying may suit you. It certainly saves shopping time and can be economical for a large family. But bulk buying is not always much cheaper than buying small cuts. You will save the retailer's profit by buying in bulk wholesale, but there are various hidden costs which you must allow for when you work out what the meat actually costs you. To make the best savings, you need to know how to assess the price you are asked, how to have the meat prepared, how long to keep it and how to use it best afterwards.

To assess the price you are asked, study the local market reports. Most prices relate to live weights and beef prices must be divided by a conversion factor to arrive at the carcase price in pence per kilogramme. The price at the wholesale carcase market should be at least $10-12\frac{1}{2}$% lower than a retail quotation, but you must expect to pay slightly more if you buy from a frozen food centre or a butcher because they have chosen the meat and to some extent have prepared it for you. You will pay slightly more than the average carcase price if you buy just a hindquarter of beef too, because it is more popular than the forequarter and so is always more expensive.

In assessing prices, remember that while lamb and pork carcasses weigh almost the same as before after they are cut up into usable joints, beef has heavy bones and a lot of useless fat and gristle; in fact, you will probably only get about 65% usable boneless beef. So, if you pay the wholesale carcase price, you must multiply it by $1\frac{1}{2}$ in order to compare it with the retail price for cuts usually sold without bone (such as steaks, skirt, topside, silverside and top rump). If you have the meat fully boned for you, the cost of doing it must be added as well.

Besides all this, you must add to the price the cost of running your freezer while the meat is in it. It will not pay you to keep the meat too long in your freezer, or to give too much freezer space to it. Bulk purchases can take up a lot of room which could be used for other goods.

Remember, too, if you are quoted a

general current carcase price to calculate for yourself what the particular carcase you are offered will cost, regardless of what any other buyers are paying. Carcasses can vary in weight a great deal and the one offered to you may be a small one.

Do not, however, buy a bigger carcase than you need just because the meat then seems cheaper. It will not be cheap if it loses quality through being frozen for longer than its high quality storage life (see page 27 below). Pork, in particular, may deteriorate and wastage of this kind is expensive. It is tempting, but not wise, to buy more meat than you will need during the coming three to four months.

A whole carcase, or a side or quarter must usually be cut up by a butcher so that you can handle it in your kitchen. You may be able to have the butchering done at the market; otherwise, have the meat transported to a butcher with whom you have made an arrangement ahead.

If you wish, you can have the meat fully prepared, that is dressed as ready-to-use joints and pieces, but this is the most expensive way to have it butchered, since the butcher's skilled labour is costly. It is a good deal cheaper to have the meat cut into the principal joints only. (See pages 7 and 8). This takes much less of the butcher's time.

You can also choose whether to have your meat fully boned, partly boned (*ie* just with the big bones removed) or on the bone. If you buy prepared meat from a multiple or private butcher, you may be able to choose which kind to buy. If you buy 'bone-in' meat, you pay least for preparation but pay for the extra weight. The flavour of the meat will be better when cooked; but against that it will take up valuable space in your freezer and you will be freezing a lot of bone which will not be eaten. For most people, having the meat partly boned is often the best choice; they pay a price between that for bone-in meat and boneless meat and can get most of the advantages of both.

When you calculate which way to buy meat in bulk, remember that almost any bulk meat purchase involves hidden personal costs at home. Usually, you must cut the meat into family-sized portions for the freezer yourself (use the same tools as for cutting up meat for pickling listed on page 12 above). You must have the time and buy the the materials, for a fairly long bout of packaging, labelling and stacking in the freezer. You must then be prepared to keep a close eye on your freezer diary, so that you do not leave the meat frozen too long and risk spoiling it. All these tasks cost you time and energy, besides the cash you must spend on knives, clingfilm, labels and so on. These charges must be added to the cost of the meat itself.

Whether, given these costs, bulk buying is worthwhile for you will depend to quite a large extent on your personal circumstances; for instance, on whether you go out to work, have access to a helpful wholesaler and butcher, possess a large freezer and have time to study the local market reports. The crucial thing, however, if you want to make bulk buying worthwhile, is to be able to use all the meat to good advantage. You need to know in advance acceptable ways to serve every bit of even the less usual offals and ways to use the bones, fat and trimmings as well.

Sample List of Purchases

Item	Quantity in each pack	Number of packs	Price per lb./ total cost	Date required
Lamb chops	4	4		June
Rump steak	½ kg/1 lb.	2		June
Mince	½ kg/1 lb.	4		June
Pork spare-ribs	1 kg/2½ lb.	2		July
Legs of lamb	1½–2 kg/3½–4 lb.	1		July
Pig's kidneys	2 kg/4 lb.	1		July
Topside	1½–2 kg/3½–4 lb.	1		August
Stewing beef	1½ kg/3½ lb.	2		August
Breast of lamb	¾ kg/2 lb.	1		August

Ordered March 31st

Buying to a Schedule

A practical way to buy meat for the freezer, if you do not want to get it in bulk, is to buy on a pre-set plan or schedule. This way of buying gives you more freedom of action to choose what you and your family like and is especially useful if your freezer space is limited.

The best way to organise it is to plan what you will need for two to four months ahead, taking into account that prices will change with the seasons (page 22 above). Make a list of the items you want, like the sample one below and how you want them packed. Then show it to a butcher or other supplier and ask him how he will charge for the frozen goods and what discounts he will allow you for the firm, ahead-of-date order.

Choose a quiet moment to discuss your proposed purchases with your supplier, as he cannot pay attention to them when he is busy. He will either quote you special prices, or, more likely suggest charging you the retail price at the time of purchase less a discount of 5–10%, depending on the size of your order. You may be able to get an even bigger discount if you take the meat before it is frozen and then package, label and freeze it yourself.

Like buying in bulk, buying to a plan is more economical than simply buying meat for the freezer when you happen to see a special offer or want to stock up. You can buy meat when it is cheapest, pack it away with the types which you will need first nearest to hand and have it processed (or do it yourself) in the way you like or find most economical.

You will only make really worthwhile savings, however, if you know exactly what you want to buy and why. For one thing, this will make your supplier respect your judgement and try to meet your wishes.

As with any kind of food, you must also know how to freeze and thaw your meat correctly, as well as how to make appetising use of the less popular (and therefore cheaper) cuts.

Knowing your meat

It would take a large volume to describe all the finer points to look for when assessing the quality of meat for the freezer. This book cannot even describe all the different cuts you may find, although the standard British cuts illustrated on pages 7–8 will help you. Before you buy for the freezer, make sure you know them well and if possible get hold of a detailed book on the subject and study it. It will save you money later on. Study the following general points on quality too.

1. Beef and lamb should have been hung, beef for 10–14 days, lamb for 7–8 days (in winter only). Fat meat can be hung longer than lean since it discolours less. All cut surfaces discolour, so meat should have been hung as a whole carcase, side or quarter, not in joints. Hung meat takes less time to cook than fresh meat and has a better flavour.

2. The spinal cord should have been removed since it spoils quickly, may smell and can cause unpleasant cross-flavours in the freezer.

3. There should be no projecting jagged bones on a joint.

4. A lamb carcase of more than 18 kg (40 lb.) is likely to be over-fat. Any fat on it should be cream-coloured and show the lean meat beneath. A mutton carcase, if well hung, can be a good buy since it has more flavour than the younger animal; it is, of course, heavier, weighing up to 23 kg (50 lb.). The usual cuts of lamb or mutton bought for freezing are legs, loin and chump chops, shoulders, best end of neck, middle neck and breast.

5. A young pork carcase can weigh up to 45 kg (100 lb.). It will have a good flavour although it may be indigestible. Do not discard the trotters, knuckle, flank and head from your purchase; you will pay a lot more per pound if you do and some delicious dishes can be made with these unpromising bits. Some examples are given on pages 43–45. Apart from these, the most usual cuts bought are legs, the middle, loin, shoulder, spare ribs, hand and belly.

An older or deliberately fattened pig known as a 'heavy hog' can weigh up to 50 kg (110 lb.). The flavour is good and the meat is more digestible than that of the younger 'porker'. Heavy hogs are usually sold to bacon factories; but if for some reason they are not used, the less choice parts are made into sausages and pies and the choicer cuts are sometimes available for sale. If you can buy them, do so. They are sold by such names as long cut or short cut hogmeat, long cut or short cut griskin, butts, tenderloin and American spare ribs or pork rib bones.

6. A hindquarter of beef should give the following principle cuts: Top-bit (which contains the thigh-bone); leg; topside, which is an overrated joint, being highly priced yet not first-class for roasting; top rump; aitchbone (have the bone itself removed); silverside (which should be boiled or braised slowly rather than roasted); rump; loin and flank, comprising the goose skirt

(for stewing) and the sirloin and sundry steaks. A fore-quarter of beef provides the clod, sticking and shin; brisket; fore-ribs; leg of mutton cut; chuck joint and sundry steaks. Besides these joints of beef, you will also get a good deal of meat for mince.

7. Weaned veal calves of two to four months old are expensive because they can be kept for beef or as milk producers. They weigh up to 45 kg (100 lb.). Dutch milk-fed calves with their whiter flesh are even more expensive than British meat, but the quarters come neatly cut with little waste, ready for packaging and so may be a good buy. A side of veal gives the haunch, comprising loin, knuckle and leg, the breast and shoulder, scrag and middle neck.

8. Bobby calves are unweaned calves sold within two to three weeks of birth. Usually small and lean, they are often quite cheap; and they make nourishing stews, blanquettes, jellies and soups, especially good and digestible for children and old people. They are cut in the same way as the older veal calves.

As well as joints, carcasses provide liver, heart, skirt (beef), melts (beef, pigs), tails, tongues, tripe, sweetbreads (lambs, calves), caul fat and chitterlings (pigs), feet (calves, pigs), and brains. Almost all these can be preserved in one way or another and most can be frozen successfully for processing later.

Freezing the meat

When you have your principal joints and other bits and pieces, you still have to package, label and store the meat in the freezer. You must also keep a record of what is in the freezer and how long it has been there.

Good packaging is important. It prevents cross-flavouring from other foods, punctured wrappings from jagged bone ends, freezer burn (unsightly discoloured patches, although not harmful), and messy spills inside the freezer which make it less efficient.

It is also important to package the meat in usable portions. If you buy carcase or partly processed meat, you will either have to do this yourself or instruct a butcher and check his pack sizes. Even if you buy ready-frozen meat in bulk from a frozen food centre, you may still have to repackage it in smaller quantities before storing it. A huge lump of mince originally packed for commercial use will need dividing into family-sized portions, for instance.

Besides the tools mentioned on page 12, a frozen food knife with a serrated edge may be a useful item to get for this task. It will also help you to save space and money by trimming joints and pieces into neat shapes.

Packaging materials must be hygienically clean and dry. Meat is usually best packed in heavy or doubled foil sheet, in polythene sheet sealed with freezer tape, or in freezer paper. Before wrapping it, however, wipe it dry and cover any projecting bone ends with doubled foil or with newspaper. Also, separate slices of meat and any small items with sheets of Cellophane or greaseproof paper. If the meat is likely to be in the freezer for some time, wrap it in mutton-cloth before covering it with the main wrapping, as this will help to prevent freezer burn.

Cover the meat with the main wrapping closely, taking care to smooth out any air pockets so that it touches the meat everywhere. It is vital to remove all air from the package before freezing, to prevent the meat deteriorating.

Secure the wrapping with freezer tape and label the meat at once. (It is surprisingly easy to forget what is in a particular package even within a few moments.) Record the kind of meat, the cut, its weight (or number of portions or items) and the date of freezing. Note, too, the date by which the meat must be removed and eaten (see table below), and any details of ingredients or seasonings you wish to add before cooking; write the labels with a felt pen or wax crayon since most other writing fades in the freezer. Before storing the food, record the details on the label in a freezer diary or notebook to remind you when to take the meat out and use it.

It is vital before freezing meat to set the freezer to the correct temperature for rapid freezing. Follow the manufacturers' instructions carefully. You must also remember not to put more in the freezer than it can freeze at one time. If you have made a bulk purchase, this may mean chilling some of the meat packages in the refrigerator for 24 hours until the others are frozen solid. Some experts say that meat should not be frozen at home at all, because it should be frozen very quickly to keep its quality and commercial blast-freezing techniques give better results. If you have bought your meat at a food freezing centre, you will have the advantage of this; and if you buy from, or have your meat fully processed by, a butcher, he will probably do the initial freezing for you. But of course you pay for it in various ways (see page 21 above), and there is no reason why meat freezing at home should not succeed, provided you freeze it as hygienically and as quickly as you can and in small quantities at a time.

When you freeze meat bought in bulk, or different kinds, freeze offals first, then pork, veal and lamb in that order. Beef will keep best in the refrigerator so freeze it last. Try not to freeze more than 2 kg (4 lb.) meat per 30 cm cubic space (1 cubic ft.) at a time.

High quality storage life

The meat cut below	*will keep its original quality for:*
Chops and steaks	
Pork	6 months
Lamb or veal	9 months
Beef	12 months
Cubed meat	2 months
Ham in the piece	3 months
Ham, sliced	1 month
Hearts, kidneys, liver, sweetbreads, tongue	2 months
Joints	
Pork	6 months
Lamb or veal	9 months
Beef	12 months
Mince	2 months
Sausage meat and sausages	1 month
Tripe	2 months

To make sure that home-frozen meat is still fit and good to eat, be scrupulously careful to take it out and use it before the end of its high quality storage life. Here are the recommended storage times. During these periods the meat should remain the same quality as when put in the freezer; if left longer, it may deteriorate.

Thawing the meat

You may spoil your meat if you thaw it carelessly. Most meat is juicier and better flavoured if it is fully thawed before cooking, whether you take it from your own freezer or buy it when you need it. To allow time for thawing you must plan ahead and take the meat out of the freezer (or buy it), 24-48 hours before you want to cook it. The length of time needed to thaw it completely will depend on the thickness of the piece and its wrappings, on whether it contains bone and on how much fat covers it.

Thaw the meat in the refrigerator, not at room temperature. It needs to thaw slowly to prevent it being wet and soft. Meat for roasting, in particular, needs to be as dry as possible and must be fully thawed; otherwise the outside may be overcooked while the inside is still semi-raw. This is not only unattractive. It may actually be dangerous, especially in roast pork. If you have to roast a joint before it is fully thawed, wrap it in foil and allow $1\frac{1}{2}$ times its usual cooking period.

Stewing and casserole meats need not be as fully thawed as meat for roasting. Any liquid which drips from them when thawing can be added to the dish while it cooks. To speed up the thawing of meats for this moist cooking, place the pack in cold water for an hour or so before cooking.

Thin cuts, small pieces or slices can be cooked while still frozen, although they will take longer to cook than usual. If you think you may need to cook them this way, try to freeze them in foil pans or freezer-to-table glass cookware which you can put straight into the oven from the freezer, and do not layer slices with paper (see page 26 above).

Use meat immediately it has thawed. In an emergency, store it in the coldest part of the refrigerator for 18 hours, but no longer. If you cannot use it then, cook it thoroughly at high heat and store it cooked. *Never* refreeze raw meat. Refrozen raw meat may have attracted bacteria which will riot when the meat is thawed a second time. Also, the ice crystals formed during freezing will re-form and be broken down again and this process may destroy a good deal of the meat's flavour.

Cooking Frozen Meats

Most people know how to roast a joint. Frozen meat joints should, therefore, give you no trouble if you thaw them slowly and fully; but, as I have stressed, it is crucial to know good and appetising ways to use any cheaper cuts, offals and even trimmings, fat and bone if you have taken the step of freezing them and want to make really worthwhile savings.

In Sections VI-IX, therefore, this book looks at ways of cooking and processing the cheaper meats which you have preserved by freezing or other methods, so that you can use them to advantage.

Meats Sealed in Fat

In places as far apart as Africa and the Arctic, the hippo and seal show that meat remains edible for quite a long time in a coating of fat. Old-time cooks developed this method of preserving and made it one of their standard ways of keeping meat for a short time. It is the method which the French still use for their *confits* (preserves), their *rillons* (cubed cooked pork) and *rillettes* (pounded meats in fat). They also use it for the pâtés cooked in earthenware dished called terrines, lined with bacon or pork fat. Most of our own potted meats and pâtés are now sealed under a layer of clarified butter, lard or other fat too. (See pages 59–63.)

The method simply consists of burying the meat in a jacket of fat so that no disease-bearing bacteria in the air can reach it. But obviously any bacteria already in the meat must be killed or made harmless first, either by curing or cooking the meat, since they can multiply inside the fat coating.

There is usually not much point in dealing with fully cured meat by sealing in fat; it is already a long-term preserved meat and extra sealing will not make it keep longer. In fact, the fat coating may become rancid through contact with the air sooner than the cured product which it coats. But sealing in fat can certainly make cooked meats keep longer, in some cases for several weeks instead of a few days.

We do not use this method of preserving as much as we might in this country, except for potted meats. Cooked sausages (page 46) and faggots (page 57) for instance will both last for several weeks preserved in this way. Pack them into sterilised pots or deep dishes when cooked and cover them with 1 cm ($\frac{1}{2}$ in.) melted lard. When the lard hardens, seal the pots or dishes with foil. Keep in a cool dry larder. To use the goods, stand the pots or dishes in a cool oven or in simmering water until the lard melts; take out the sausages or faggots by spearing them with a skewer, drain over the pot or jar and fry as briefly and quickly as possible to heat them thoroughly to the centre and to crisp the outside.

You will find the potted meats in Section IX on pages 59–63 and pâtés and bacon-lined terrines in Section X on pages 64–72. Like French rillettes, they are all made with very finely minced or pounded meat.

To seal solid pieces of meat in fat, use the following standard recipes for making French pork rillons and *confit d'oie* (preserve of goose). This goose confit is the best known fat-sealed preserve, but other meats are just as good, treated in the same way. Use this method for dealing with the remains of a Christmas turkey, for instance.

The bird usually produces plenty of fat, too highly flavoured for everyday use, but ideal for sealing; and the flesh can be kept as a confit for several weeks instead of being used immediately after Christmas when most people do not want it. Duck, pork, or rabbit also make tasty confits.

Preserve of Goose

1 goose
Salt as required (see recipe)
A few cloves, crushed black peppercorns and garlic cloves tied in a muslin square
Goose fat and lard as required (see recipe)

Joint the bird and remove the large bones. Weigh the meat. Rub the joints thoroughly with 25 g (1 oz.) salt for each 480 g (1 lb.) meat. Refrigerate overnight.

Prepare the spices. Also have ready half the weight of the meat in fat. Use fat from the joints and make it up to the required weight with lard. Melt the fat in a large, heavy-bottomed casserole. Wipe the goose thoroughly and add it to the casserole with the bundle of spices. Cover and cook at 180°C, 350°F, Gas 4 for 2-3 hours, depending on the size of the joints. Test whether it is done by piercing a thigh with a thin, well-heated skewer. It is ready if no liquid comes out.

While cooking the meat, sterilise a china, glass or heavy earthenware pot or jar to put it in.

When the meat is cooked, discard the bundle of spices, strain a layer of fat into the jar and let it get cold and firm. Drain the remaining fat from the meat, but keep it. Remove any skin and sinews from the meat. Then pack the meat into the jar on top of the firm fat layer, without letting it touch the sides of the jar. Leave 5 cm (2 in.) headspace. Scrape any meat juices off the unused fat, or pour off the pure fat if liquid. Remelt the pure fat if required and pour it over the goose meat, covering it well. Allow it to get cold and firm. Then add another thin layer of fat to the jar, to seal it completely. When cold, seal with foil or clingfilm.

Keep in a cold place or refrigerator. Do not keep longer than six months.

To use, remove the pieces you want with a long skewer dipped in boiling water. Scrape off any fat clinging to them or toss the meat over heat (if to be eaten hot) to melt the fat. Pour the fat back into the jar, to seal in the remaining meat. Allow it to get cold and firm, then recover as before.

You can eat the pieces cold with salad, but they are better hot. Serve with mashed potatoes, or mix with other meats, chopped smoked sausages and white beans in a rich stew.

Make a preserve of turkey, duck, pork or rabbit in the same way, using turkey or duck fat as appropriate; use pure lard for pork or rabbit.

If you wish, you can use meat already cooked such as the remains of a Christmas goose or turkey with the fat it has produced. Trim the pieces neatly and remove any skin, sinews and large bones before cooking. Do not salt. Otherwise, treat like the raw meat above, but cook only until the meat is really well heated through to the very centre. The time will depend on the thickness of the pieces. Pot, keep and use as above.

Typical crock for Preserve of Goose   *John Harris*

Pork Rillons

$\frac{3}{4}$ kg (1$\frac{1}{2}$ lb.) lean fresh pork
1 kg (2 lb.) fat pork, *e.g.* belly with some hard fat included
1 × 2.5 ml spoon ($\frac{1}{2}$ teaspoon) freshly ground black pepper
3 × 2.5 ml spoon (1$\frac{1}{2}$ teaspoons) salt
1 × 2.5 ml spoon ($\frac{1}{2}$ teaspoon) crushed dried thyme
$\frac{1}{2}$ × 2.5 ml spoon ($\frac{1}{4}$ teaspoon) crushed dried sage or rosemary
1 bay leaf
1 × 15 ml spoon (1 tablespoon) finely chopped fresh parsley
8 fl. oz. (200 ml) boiling water

Sterilise attractive small glass or china pots to hold the rillons.

Cut the lean and fat pork into 5 cm (2 in.) cubes. Place in a heavy-bottomed pan with all the other ingredients. Cook over very low heat, stirring occasionally, until all the water has evaporated and the cubes are golden-brown on all sides. Turn the meat and all the fat in the pan into a sieve placed over a basin. Allow to drain very thoroughly. When all the fat has drained into the basin, chill it until it is firm.

Discard the bay leaf. Scrape off any meat juices under the fat in the basin. Melt the cleaned fat and pour a thin layer into each of the jars. Mix most of the remaining fat with the spiced cubes, reserving enough to cover the jars with a 1 cm ($\frac{1}{2}$ in.) layer. Let the layer of fat in the jars get quite firm, then pack the pork cubes on top, packing them down to squeeze out any air spaces. Level the top surfaces. Remelt the reserved fat if necessary and cover the pork cubes with a 1 cm ($\frac{1}{2}$ in.) layer of fat. When the fat is firm and cold, cover with jampot covers or lids and store in the refrigerator until required. Use within four to six weeks.

Serve as a cold hors d'oeuvre with dry, hot toast, like a pâté. But since rillons are richer than a pâté, serve sliced pickled cucumbers or another sharp pickle with them.

Spiced Meats

All cured meat is spiced. Salt and most other preserving ingredients are spices which flavour the meat as well as preventing it going bad. Spices also keep fresh raw meat untainted for a short time, without going through the process of curing it fully. This can sometimes be extremely useful if you cannot process or cook it immediately after buying it.

Marinades

The most usual way to treat fresh raw meat with spices is to steep it in a marinade. This is a seasoned liquid containing some kind of acid, which works on the game or meat placed in it; it adds flavour, softens the fibres and preserves the meat for a short while. (The word 'marinade' comes from the Spanish *marinada*, meaning to pickle.)

A marinade can be cooked or uncooked. A cooked marinade keeps better; and it can be used more than once if reboiled, although it is more often strained and used for cooking the meat. Bigger pieces of meat are usually soaked in a cooked marinade.

An uncooked marinade works on the meat more slowly but more subtly. It is better suited to small pieces of meat and poultry, such as chops and chicken legs. If it shows any sign of fermenting in hot weather, the meat can be removed and the marinade can be boiled up with a little extra wine or vinegar. When it is cool, the meat can be replaced in it without any harm.

There is no hard-and-fast rule for how long meat should be left in a marinade, either cooked or uncooked. If you just want to flavour and tenderise meat such as stewing steak, even two to three hours in a marinade will improve it. But if you want to process a large piece of meat, or a joint which you cannot use at once, you can keep it safely in a marinade for up to 48 hours in summer, two to three days in winter. Very large joints, such as venison, may benefit from being left in it even longer.

There are a great many different marinades, but the basic recipes on pages 34-35 below will cope, between them, with most kinds of meat and meat cuts. These recipes use ingredients available in most housewives' kitchens. They have the added virtue that they keep well and can be prepared ahead of time, to use when needed.

Marinating is a great help in dealing with cheaper cuts of meat since it tenderises them as well as flavouring them. It can be extremely useful too, if you buy meat in bulk for freezing or curing and lack freezer space or time to process all the meat at once. You can put some of the cheaper cuts into a marinade temporarily, until you can handle them; and you will know that they are improving in flavour and tenderness, yet are safe from taint, while you process pieces needing more urgent attention.

It may even be a good idea to prepare a marinade deliberately before taking delivery of an unpackaged bulk order of meat in order to give these cheaper cuts at least a short soaking while you handle the rest. It will improve them in any case; and it will certainly be useful if any unexpected interruption occurs while you are processing the order.

It may even save the day if, for any reason, a power failure puts the refrigerator or home freezer out of action for a time.

Marinating is a helpful kitchen technique to use regularly if you live in a rural area or cannot shop often for some other reason. It is valuable too, for anyone without a refrigerator.

It can also be extremely useful for keeping meat over a long holiday week-end. You can often get a wider choice of meat, and get it more cheaply, if you buy a few days ahead of the holiday; and you can do so securely if you can marinate the meat for a day or two.

You need never make a large quantity of a marinade, provided you turn the meat over in the liquid from time to time and baste it often. Baste with a wooden spoon and use a glazed earthenware, stoneware or glass container; the marinade should not come into contact with metal. It should also be sheltered from bright sunlight.

Uncooked Marinade 1
(for large or small cuts of red meat)

1 medium-sized onion, peeled and sliced
1 medium-sized carrot, chopped
1 stick celery, chopped
1 clove garlic, crushed
6–10 parsley stalks, chopped
1 × 5 ml spoon (1 teaspoon) dried thyme
1 bay leaf
6–8 peppercorns
1 clove
1 × 2.5 ml spoon ($\frac{1}{2}$ teaspoon) ground coriander (for game only)
1 × 2.5 ml spoon ($\frac{1}{2}$ teaspoon) juniper berries (for game only)
Salt and pepper to taste
250 ml ($\frac{1}{2}$ pint) medium dry red wine
125 ml ($\frac{1}{4}$ pint) water
125 ml ($\frac{1}{4}$ pint) salad oil

Tie the chopped vegetables, herbs and spices very loosely in a cloth so that it will lie flat like a cushion. Use butter-muslin or cheese-cloth. Lay the 'cushion' in a dish, put in the meat, and pour the liquids over. Marinate the meat for at least six hours, turning the whole contents of the dish over occasionally, so that the 'cushion' is sometimes on top of the meat. Keep in a cool place.
Note: This is a good marinade for tough stewing cuts.

Uncooked Marinade 2
(for large or small cuts of white meat)

Salt and pepper to taste
1 onion peeled and finely sliced
6–10 parsley stalks, chopped
1 × 5 ml spoon (1 teaspoon) dried
 thyme
1 bay leaf
1 clove garlic, crushed (optional)
Juice of 1 lemon
50 ml (2 fl. oz.) salad oil
250 ml ($\frac{1}{2}$ pint) mixed white wine and
 water or dry cider and water

Use in the same way as the previous marinade.

Cooked Marinade for Red or White Meat

1 carrot, about 50 g (2 oz.)
1 onion, chopped, about 75 g (3 oz.)
1 × 15 ml spoon (1 tablespoon)
 chopped shallot (optional)
25 g (1 oz.) chopped celery
1 clove garlic, crushed
1 × 5 ml spoon (1 teaspoon)
 chopped fresh parsley
Pinch of dried thyme
1 bay leaf
6–8 peppercorns
1 clove
6 coriander seeds (for strong game
 meat only)
6 juniper berries (for strong game
 meat only)
Salt and pepper to taste
125 ml ($\frac{1}{4}$ pint) salad oil
500 ml (1 pint) medium dry red or
 white wine, depending on the kind
 of meat (see note)
100 ml (4 fl. oz.) wine vinegar

Slice the carrot thinly. Simmer all the vegetables and herbs in the oil gently until lightly browned. Add the wine and vinegar and simmer for 20–30 minutes. Allow to cool completely. Soak the meat in the mixture for at least six hours, turning it over from time to time. Keep in a cool place. When ready to use, wipe the meat thoroughly to dry it. If braising, strain the marinade and include in the cooking liquid.

Notes:
1. As a rule, use red wine for red meats, white wine for veal, pork or chicken. If used for cooking the meat, red wine gives a richer sauce; white wine rather accentuates the meat's own flavour.
2. This marinade can be kept for quite a long time if it is reboiled from time to time. Boil it every other day in hot weather, twice a week in cold weather. Each time, add a little extra wine and vinegar.

Variations:

To vary the flavourings, add a pinch of one or more of these: Dried rosemary; dried sweet basil, oregano; grated orange or lemon peel; grated nutmeg; ground cinnamon.

Light Salting

Marinating is not the only method used for spicing meat, to preserve it for a short time. Light salting (as recommended for goose in Chapter III) is one alternative often used, especially for pig meats. The meat can be dry salted, or can be put into a brine bath like the pig's chap below. Some cookery experts claim that all pork is improved by 12–24 hours' salting before cooking,

35

Collared Breast of Lamb *John Harris*

even meat for roasting. Certainly cheaper cuts and the lesser variety meats such as trotters and trimmings, gain flavour if salted. So, except in very hot weather, they can be bought a day ahead of being needed, to allow time for it.

Dry Spicing

Instead of being salted, most sausage and other minced meat mixtures have dry spices mixed into them, to preserve them. Potted meats and pâtés often have both spices and a good deal of fat mixed into them. The sausage mixtures are described on pages 54–58 and the pâtés and similar mixtures on pages 64–72.

Collared Meats

One other way of spicing meat is to collar it before cooking it. This is an old, uniquely English process, not developed in quite the same way anywhere else. It is an excellent method of dealing with cheaper, less attractive cuts such as breast of lamb and veal or brisket of beef, if you have not got time to pickle them.

Collaring was developed in the 17th century. It got the name because it was used for long flat pieces of meat which were spiced and then rolled up like the high neckbands fashionable for men at that time. Sometimes, in past days, the meat was salted or pickled before or after being collared and cooked, but the traditional recipes below do not include any curing.

Cooked collared meats can be kept for several days before being eaten. One of their attractive features is that, when they are sliced, the herbs and spices spread on the meat before rolling it show as circles of bright colour in the roll. They give it an air of gaiety without any great expense. The French use pistachio nuts for similar meat products and for pâtés, but they are much more costly.

Pig's Chap or Trotters

½ pig's jaw and cheek as for Bath Chap
Spiced brine made with:
    2½ litres (5 pints) water
    300 g (12 oz.) coarse salt
    300 g (12 oz.) brown sugar
    1 × 2.5 ml spoon (½ teaspoon) saltpetre
    1 × 5 ml spoon (1 teaspoon) juniper berries
    Fragments of whole nutmeg, to taste
    1 bay leaf
    2-3 sprigs dried thyme
    1 × 5 ml spoon (1 teaspoon) peppercorns
    3 cloves

Stock made with:
    2 onions, sliced (not skinned)
    2 carrots, coarsely chopped
    2 leeks (optional)
    1 clove garlic
    2 bay leaves
    6 parsley stalks
    2-3 sprigs dried thyme or basil
    6-8 peppercorns
    2 × 15 ml spoons (2 tablespoons) wine vinegar
    500 ml (1 pint) white wine or dry cider
    Water (see recipe)
Toasted breadcrumbs

This is a domestic version of the well-known Bath Chap, which is cured and smoked like a ham before boiling. It will not keep as long as a fully-cured Chap, but it is a good way to make this economy pig meat a presentable dish.

Ask the butcher to cut off the jaw as for a Bath Chap. Prepare the brine. Bring the first four ingredients to the boil in a large pan, while tying the rest in a square of muslin or thin cotton. Skim the boiling brine, remove from the heat, add the spice bag and allow the brine to cool completely before adding the meat. Steep the meat in the brine for 48-72 hours. (Brine several meats at the same time. The Chap is hardly worth doing by itself.)

When you remove the meat, place it in a pan of fresh cold water. Bring it to the boil. Remove the meat and wipe it well to remove as much salt as possible.

Put all the stock ingredients into a large pan. Tie the herbs and spices in a thin cloth bag first. Tie the meat tightly in buttermuslin too, to keep it in shape. Add enough water to cover the meat. Bring gently to the boil, cover the pan securely, then reduce the heat to as low as possible so that the stock barely simmers. Simmer for three to four hours.

Remove the meat and let it cool slightly. When it is just cool enough to handle, remove the bones carefully. Form into the shape of a cone cut in half lengthways. Allow to cool to tepid, then remove the skin. Allow to cool completely, with a weight on top. Coat with toasted breadcrumbs and serve cold, thinly sliced.

Note: Trotters are good treated in the same way, completed by rolling in toasted crumbs. They are usually eaten reheated, *e.g.* lightly grilled.

Collared Breast of Veal

1 breast of veal
The following to taste:
   grated nutmeg
   ground mace
   salt and pepper
   finely chopped fresh or dried mixed herbs
   chopped parsley
   grated lemon peel
½ slice white bread, crumbled
2 × 15 ml spoons (2 tablespoons) clean dripping
Flour for dredging
25–50 g (1–2 oz.) unsalted butter

For the sauce:

1 beef stock cube
Flour as required
1 onion, shredded
Bundle of fresh herbs
1 slice toast
The veal bones

Bone the meat carefully without cutting the flesh right through. Place the bones in a stewpan to make the sauce (if eating the veal hot). Lay the veal flat, skin side down and sprinkle well with the nutmeg, mace, salt and pepper, herbs, parsley, grated lemon peel and breadcrumbs. Roll up tightly, secure with skewers or tape and wrap in foil greased with the dripping.

   Pot-roast at 170°C, 325°F, Gas 3 for 1–1½ hours or until the meat is tender, 20 minutes before the end of the cooking time, remove the pan lid and foil, dredge with the flour, baste with the butter and allow to brown.

   Keep warm while you make the sauce, if to be eaten hot. Otherwise allow to cool completely, wrap closely in clingfilm and keep in a cold place or in the refrigerator until required.

To make the sauce:

Pour enough boiling water on the stock cube to make double strength stock. Use a little stock to make 1 × 15 ml spoon (about) (1 tablespoon) flour into a paste. Add the paste to the liquid stock a little at a time, over gentle heat. Stir in and continue stirring until the sauce comes to the boil and thickens. Add the shredded onion, herb bundle, toast and bones and simmer until the onion is tender. Add a little water if the sauce gets too thick. Test the seasoning and strain the sauce before serving.

Collared Breast of Lamb

2 breasts of lamb
1 egg yolk
1 × 10 ml spoon (2 teaspoons) grated lemon peel
Salt and pepper to taste
Pinch of grated nutmeg
1 small bottle of capers, drained
6 anchovy fillets
2 × 15 ml spoons (2 tablespoons) chopped parsley
1 × 5 ml spoon (1 teaspoon) dried marjoram
1 × 5 ml spoon (1 teaspoon) dried chives
100 g (4 oz.) soft white breadcrumbs
750 ml (1½ pints) chicken stock from cubes

Lay the breasts flat on a table, bones uppermost. Remove the bones without cutting the flesh right through. Place the bones in a large casserole with a lid. Brush the boned sides of the breasts with the egg yolk, then with the peel, salt and pepper and nutmeg. Chop the capers and anchovy fillets and place in a small basin. Mix in the parsley, herbs and breadcrumbs and moisten with a little of the stock. Spread this mixture over the breasts. Overlap the breasts a little and roll them up into a single tight roll. Secure with string. Wrap tightly in a piece of buttermuslin and tie again, leaving long ends of string which will hang over the side of the casserole and help you take the meat out when it is done. Lay the meat in the casserole on the bones. Add the stock and simmer for 2½–3 hours. Eat hot with white beans or serve cold with a salad which includes fresh chopped mint.

Collared Beef

1½–2 kg (3–4 lb.) fresh brisket or silverside
Salt and pepper to taste
A sprinkling each of ground cloves, mace and ginger
A few drops of cochineal
½ × 2.5 ml spoon each of:
   allspice
   pepper
   dried sage
   dried thyme
1 bay leaf

Lay the meat out in a flat strip; if thick, cut it through horizontally and lay the two pieces end to end. Season it with salt and pepper and with the sprinkling spices and rub it with the cochineal. Roll it up tightly and tie it with string at each end and in the middle. Wrap it in one thickness of buttermuslin. Place it in a stewpan with water to cover and bring gently to the boil. Remove any scum, add the remaining herbs and simmer gently for three to four hours or until tender. Serve hot with pease pudding and pickles or cold with a salad which includes fresh herbs and pickled mushrooms or gherkins.

Brawns and Head Cheese

Brawn is a very old English dish. We usually think of it as chopped pork set in jelly. But there is a solid type of brawn as well as a jellied one. The word 'brawn' simply means the flesh or muscle of a boar (today, a pig) which has been collared, boiled and pickled, or potted.

There are several ways to make both types of brawn. For instance, other kinds of meat besides pork can be used and the lesser ingredients can vary too; often, they depend on where the brawn is made, since different recipes have been developed in various parts of the country.

In the past, solid brawns were sometimes made with long strips of meat from the pig's flank (belly). They were rolled up and collared like the meats on pages 38-39 and were then kept in pickle until wanted. You can use the Collared Beef recipe on page 39 to make a solid pork brawn like those old ones, using belly of pork if you wish, instead of brisket or silverside. It is good to eat cold when thinly sliced. Remember, though, that the collared meat recipes above do not include full curing, either before or after boiling, so do not try to keep the brawn too long.

Another old way of making a solid brawn used the fattier meat from the pig's cheeks and head. It is still popular and the basic method has not changed at all. The pig's head is lightly cured like the Pig's Chap (page 37) or is fully cured with the rest of the carcase when preparing bacon and hams. But instead of being dried and stored (pages 9-17), it is boiled long and slowly, usually with the tail, tongue, trotters and other trimmings and sometimes with other meats. (The brains are used separately.) When the meat is very tender, the bones are taken out. The meat is chopped and with the tongue, whole or sliced, is rolled up tightly in the skin of the head, to make a meat roll shaped like the collared meats. Two modern recipes for solid brawns using head meat are given on pages 43-44.

Pork brawn with jelly was also developed long ago to use up the fattier, small pieces of meat from the pig's head. Like the solid brawns, the general method of making it used today, even by commercial suppliers, is still the one used in the past. Again, the pig's head (or hock) and other scraps are fully or partly cured in brine, then boiled long and slowly. When very tender, the bones are removed and put back in the cooking liquid, which is left to simmer to make a rich stock which will 'jelly' when cold. Sometimes, especially if the meat has only been lightly cured, pickling spices and wine are added to the stock. When it is ready, the boned meat is chopped or cubed and is put into a mould. The stock is then mixed with the meat or is poured over it, to set it.

How much stock is added to the meat depends on the particular recipe being used. You can pack the meat down firmly in the mould, leaving hardly any space for liquid; or simply jumble the meat cubes in the mould with plenty of spaces between them and pour a lot of stock over them. In this case, there is almost as much jelly as meat when the brawn is turned out and sliced.

You can often see this if you buy commercially-made brawn. A piece from one shop may be a close-set mass of meat almost like a pâté; another, from a different supplier, even in the same area, may be a rich brown jelly with cubes of meat set in it. Yet another may be almost solid inside but be cased in jelly like a galantine. (A galantine is really just a solid brawn with a coating of jelly.)

The colour of the brawn can vary too. Brawn which is lightly salted but not cured will be greyish, not pink. The jelly may be pale if made with pork bones alone, but will be a rich glossy brown if onions in their skins or beef bones have been boiled in it, or if pickling vinegar, treacle or brown sugar have been used to make it a kind of 'jellied pickle'.

You can make whatever kind of jellied brawn you prefer at home. A standard traditional recipe is given on page 44 below, including some of the other meats you can add. All three brawn recipes include too, directions for using some of the other pig 'bits and pieces' which you may have in hand. Making brawn is one good way of dealing with them, if, for instance, you buy a whole carcase or bulk supply for freezing, but only want to give freezer-space to the better cuts. (A pig's head takes a lot of space.)

One warning. The jellied stock will not preserve the meat. In fact it may make it hazardous to keep for long, unless pickling spices have been added to the stock.

Various extra flavourings are nearly always added to pork brawn, whether pickled or not. It needs spicing well, since it has not much flavour alone. Nutmeg and sage have always been the favourite flavourings; old-time housewives used the bits of whole nutmegs too small for grating. Parsley, thyme, bay and lemon peel are also common flavourings. But many others are popular too. So when you make brawn, take the opportunity to use up the last remains of dried herbs and spices in your jars, so that you can replace them with clean new stock.

Besides spices, other ingredients may be used in the brawn. Some recipes include hard-boiled eggs, or manufacture jellied stock with gelatine. But other meat is the most important extra ingredient certainly in a pig's head brawn. The meat may be other parts of the pig, such as the hock and trotters, included because it is a convenient way to use them. Meat from other animals is most often added in order to mix dry meat with the fatty pig's head meat—which improves both—or to add flavour. An old boiling hen is often added for these reasons, making a 'Chicken Brawn'. (A roasting chicken has not enough flavour.) In some places, an ox-tongue is traditionally placed in the centre of a large solid brawn. But

Brawn

other flesh meat can be used equally well, provided it is not very fat, is well flavoured and not likely to taint. Only avoid using lamb; it does not mix well with pork and is better used for other products.

Lamb alone, however, makes a well-liked Sheep's Head Brawn; it is made in exactly the same way as the pig's head brawn.

In Scotland, brawn is called Potted Head and is made from half an ox-head. But a more modest brawn can be made using shin of beef and a calf's foot. Rabbit Brawn, still more modest, is one of the most widely made brawns.

Pork Brawn has as many names as it has ingredients. Some indicate the main flavour; Sage Brawn, for instance, or Clove Brawn. Others are place-names and only mean that the county or city has created its own, usually minor, variation of the basic pig's head brawn. Other local and some general names are more confusing. Potted Head can mean a *pig's* head brawn in some parts of Scotland, while North Country people call their brawn Pressed Pig's Cheek. The most common second name for pig's head brawn however, is Head Cheese. 'Cheese' means clots or shreds pressed to the same consistency as cheese; so Head Cheese is a name given to any chopped meat brawn, drained and packed down solidly into a mould like a cheese.

Whatever you call it, the value of a brawn is that it makes acceptable so many less attractive pig meats which you may have to take if you buy in bulk, or which you buy because they are cheap. Brawn is only a true meat preserve if it is made with well pickled meat and is solid; otherwise it must be eaten within a day or two of being made. But making brawn releases freezer space for longer-term meat preserving of better cuts and saves you wasting money by having to discard, or pass by, the less popular ones.

Solid Pig's Head Brawn 1

1 pig's head, ears, tongue and feet
Coarse and fine salt as required, for rubbing
$\frac{1}{2}$ × 2.5 ml spoon ($\frac{1}{4}$ teaspoon) saltpetre
150 g (6 oz.) granulated sugar
150 g (6 oz.) coarse salt
Vinegar as required
1 × 10 ml spoon (2 teaspoons) grated nutmeg or ground mace
Cayenne pepper to taste
A few grains ground cloves
Pickling spice to taste
Bundle of fresh herbs
1 onion, quartered
1–2 sticks celery, coarsely chopped
Finely toasted breadcrumbs

Ask the butcher to split the head in two and to bone it. At home, put aside the brains for use separately, remove the ears and tongue and any fragments of loose bone. Rub the inside of the head, the ears, tongue and feet well with mixed salt. Leave to drain (*e.g.* on a sink draining board) overnight.

Wipe off the brine and rub the meats well with the saltpetre and sugar. Leave for eight to nine hours, then rub with the 150 g (6 oz.) salt. Place in a large basin or trough. Leave overnight, then cover the meats with vinegar. Leave them in this pickle for four to five days, turning them over once daily.

To make the brawn, first wash off the pickle. Place the ears, tongue and feet in cold fresh water, bring to the boil and simmer gently until the bones will slip out of the feet easily. While simmering, lay the head flat, skin side down, on a board. Remove the tongue from the water, skin and slice it. Cut the pork meat carefully off the skin and re-arrange it in alternate fat and lean layers, using the sliced tongue as lean meat. (Add the sliced ears and meat from the feet if you wish, or keep for another dish.) Sprinkle each layer of meat with the spices.

Roll up the layered meat in a stout cloth and tie it securely with tape or string. Place it in the pot in which the ears and feet have been boiling and add the herb bundle and vegetables. Simmer for four to five hours. Leave to cool in the broth. When cool, place in a basin with a weight on top until quite cold. Remove the cloth and cover with toasted crumbs. Slice thinly and serve cold.

Solid Pig's Head Brawn 2
(with Ox-Tongue)

1 lightly pickled pig's head, ears, tongue and feet
Coarse salt as required
200 g ($\frac{1}{2}$ lb.) pork sausages
1 lightly pickled ox-tongue
Salt and pepper to taste
Dried sage to taste

Cut the pig's head in half and soak it in cold water overnight. Remove the ears and tongue. Rub all the pieces well with salt and leave to drain for 24 hours.

Wipe the head well and put it alone into cold fresh water. Bring to the boil and simmer gently for six hours. Remove the head from the liquid and place the liquid aside for later use. Let the head get quite cold, then remove all the bones. Return it to the liquid with the ears, tongue, feet and the ox-tongue. Simmer for $2\frac{1}{2}$ hours. Add the sausages and simmer $\frac{1}{2}$ hour longer.

Remove all the meat from the liquid and let it cool slightly. Trim and skin the ox-tongue. Remove all bones and gristle from the other meat. Chop all the pig meat including the sausages into small pieces, seasoning thoroughly with salt, pepper and sage. Lay the ox-tongue in the centre of the brawn mould and pack the chopped meat tightly round it. Allow to stand overnight in a cold place with a heavy weight on top.
Note: The brawn mould should be a large oblong loaf tin in which the ox-tongue can be laid flat, upside-down.

Standard Jellied Brawn

$\frac{1}{2}$ salted or lightly pickled pig's head, ears and tongue
2 pig's trotters or 1 dressed cow heel
The pig's heart (optional)
1 old boiling fowl (optional)
2 onions, halved but not skinned
Bundle of fresh herbs, including a strip of lemon peel, 12 peppercorns, 1 blade of mace and 6 cloves
125 ml ($\frac{1}{4}$ pint) white wine, or wine vinegar
Brown sugar to taste

Trim the head, removing the ears and tongue. Clean and trim the trotters or cow heel. Soak all the meat except the heart and fowl (if used) in cold water for six hours. Place in cold fresh water, add the onions and bring to the boil. Simmer gently for four hours or until the bones slip out easily, adding the heart and fowl for the last $1\frac{1}{2}$–2 hours, the time depending on the bird's size.

Remove from the heat and allow to cool slightly. Take out all the bones including the chicken bones and return them to the cooking liquid which should still be simmering. Reduce the stock to about half its volume, then strain it and return it to the pan.

While it reduces, dice or cube all the meat, using larger cubes if you want a brawn with plenty of jelly. Return the meat to the stock with the bundle of herbs, the wine or vinegar and a little brown sugar to taste. Simmer gently until slightly reduced and thoroughly mixed. Strain and discard the herb and spice bundle; place the meat in bowls or moulds, either packed down or loosely as you prefer (see page 41 above). Pour the strained stock over

and leave in a cool place to set.
Notes:
1. You can vary the flavouring by using orange peel, allspice berries, fragments of nutmeg, or a sprig of sage in place of some of the herbs and spices above.
2. Sliced hard boiled eggs can be placed in the mould under the meat and will look decorative when the jellied meat is turned out.

Rabbit or Beef Brawn
(uncured)

1 rabbit and 2 pig's trotters or
200 g ($\frac{1}{2}$ lb.) boneless shin of beef
    and 1 calf's foot
Bundle of herbs including some
    spices, *e.g.* 8 peppercorns, 1 blade
    of mace and 2 cloves
1 onion
1 bay leaf

Joint the rabbit if used and leave in cold salted water. Clean all the other meat. Cube the beef.

Place the trotters or calf's foot into cold water, bring to the boil and remove as the scum rises. Place in clean water. Add the beef if used. Bring to the boil and skim well. Add the bundle of herbs and spices and the onion and bay leaf. Simmer for four hours. Add the rabbit, if used, after two hours.

When the meats are very tender, remove them, strain the stock and replace it in the pan. Discard the herb and spice bundle, onion and bay leaf. Remove all bones from the meat and chop or cube the meat as desired. Return it to the strained stock and bring to the boil. Strain again. Pack the meat into moulds pressing down or jumbling the cubes loosely as you wish. Pour enough strained stock over the meat to cover it. Leave to set. Turn out and serve thinly sliced.

Sausages

Making sausages at home is one of the most practical ways to use cheaper bits and pieces of meat. You can adapt the shape, size and contents to suit almost any meat cut and any kind of meal.

Sausages can be made from fresh, salted, pickled, smoked or cured meat or a mixture of meats; almost any kind of meat in fact, except frozen. It can be chopped coarsely or finely, or minced, as you prefer. Quite a lot of fat is always included. Most sausages contain herbs or spices for flavouring too and some have flour, breadcrumbs or other starchy food added.

A basic sausage meat recipe is given on page 54 below. To make the mixture into sausages, one usually stuffs it into a skin or casing. This may be a natural skin from the animal or an edible synthetic covering made of cellulose. Instructions for handling it are on page 51.

Various kinds of sausages, both cased and uncased, have been popular worldwide, since ancient times. But sausages as we know them were really first created by the Romans, in order to preserve and use pig meats, which went off quickly in their warm Italian climate. Almost all sausages have been made mainly of pork ever since. But even the Romans added other meats, and most kinds of sausages developed by other peoples have included beef, veal, liver or other offals, or even blood, among their ingredients.

Today, beef sausages without any pork at all are common. Beefburgers are really just skinless flat sausages with a high (80%) meat content too, while Jewish meat balls are round ones. As for mixed meat sausages, hundreds of recipes exist. A few examples are given on pages 50–56 below, but there are many more types: Black puddings or blood sausages and other boiling rings, for instance!

Various inner membranes of the animal are used as natural coverings for the different meat mixtures. The large and small intestines are the commonest ones, although the fatty caul, a membrane enclosing the animal's organs, is also used a good deal; traditionally, for instance, to cover faggots (page 53). Tripe makes another natural covering; haggis is really an old type of sausage stuffed into a tripe bag made from a sheep's stomach, or (today) a cellulose synthetic covering. (A pig's tripe is eaten as a dish, rather than used as a covering, far more widely on the Continent than here.)

If you want to make conventional sausages in skins, you will probably have to find a kindly butcher who makes his own sausages and will supply a small quantity of casings from his stock. The suppliers' bundles are usually too big for ordinary household use, whether the casings are natural or synthetic. Synthetic casings are easier to handle. Natural casings are likely to come processed and salted. They must be rinsed thoroughly in fresh lukewarm water, then rinsed in cold; any not used must be resalted. If you use casings not yet processed (*e.g.* ones from a home-killed carcase), they must first be soaked in water for 12–24 hours, depending on the weather; then they must be cleaned of fat, turned inside out (easier to do if you cut them into shortish lengths first) and cleaned of their contents and mucous lining. It is a messy job, best done over a sink or bath. Even then they are not ready. They need to be simmered for an hour or so in fresh water, and cooled thoroughly before use.

Given these problems, it may well be easier to use caul fat if you want to use a home-processed skin. It will be easier to obtain, at least at some seasons. A helpful butcher may order it for you, or you can get it at a wholesale carcase market if you buy in bulk there. It is easy to use. Soak it in tepid water with 1 × 15 ml spoon (1 tablespoon) vinegar per litre ($1\frac{1}{4}$ quarts) of water. Then cut it into the shapes and sizes you want, and roll the meat stuffing in it, overlapping the edges well. With its veining of fat, it may look more attractive than a conventional skin, and will give the sausages more flavour when they are cooked. Details of how to use it are on page 51 below.

Some people feel squeamish about using natural casings. But the only real difference between a natural sausage casing and skin used for greasing a pancake pan is that it comes from inside the animal instead of outside. It may be comforting to think that it is really no different from bacon rind or the crackling on roast pork.

However, if you find casings unattractive or difficult to get hold of, a good alternative is to make the kinds of sausages which do not need skins at all. These are described on page 51.

Whether you use a casing or not will depend on how you want to process and use the sausages. One of their great merits is that they can be prepared and used in so many different ways. They can be eaten fresh, or can be dried or smoked for storage if cased. Most are cooked before being eaten; they can be boiled (if in skins), baked, grilled or fried. Choose the method which suits your convenience or the dish you want to make, taking into account the size and shape of the sausages.

Most sausages are shaped as cylindrical rolls which can be cooked whole, cut into chunks or sliced, depending on their size and the use they will be put to. But sausages such as the French *crépinettes* are flattish ovals, while faggots (page 57) and similar Continental sausages are like brown golf-balls. Both these types are usually served whole. So are small sausages shaped like rolls up to 75–100 g (3–4 oz.) in weight. Larger, fatter sausages have to be cut in pieces to make individual servings. Smoked sausages eaten raw as an hors d'oeuvre are sliced as thinly as possible, and some liver sausage mixtures are soft enough to spread like pâté. Other sausages can be cut into short lengths, or are cubed for cooked dishes and salads.

The parts of a pig or other animal which you can use for sausages vary as much as the shapes you can make.

Pork or ordinary sausage meat can come from almost any part of the pig not needed for other dishes; from the head, hand and spring, belly and so on, down to the tail. The less attractive interior pieces such as the lights and spleen can also be used for some sausages (page 57). Cheaper beef cuts such as skirt, or neck and clod trimmings, find a home in beef sausages, while similar cheap meat from a lamb's carcase can be used for Scandinavian salami or for the mutton boerewors on page 58. The dryer parts of game meat can be used to good effect in game sausages because the fat added to the meat mixture offsets its dryness. Almost any animal's liver can be used for liver sausage, although pig's liver is most often used.

Sausages are, therefore, an excellent way to use cheaper cuts and varieties of meat which give good nourishment but lack charm otherwise. If they look undesirable or have little flavour, they can 'disappear' in a chopped sausage meat mixture which includes strong spice flavourings. Again, a family which prefers roasts will usually accept fried or grilled sausages made from meat which would otherwise have to be stewed.

Making sausages can therefore save you giving freezer-space to variety meats if you get them in a bulk order or mixed pack, or as an economy measure. Home-made sausages save refrigerator space too, being compact; and they are an excellent emergency stand-by to have in stock, if frozen or sealed in fat after being made (page 29). Frozen fresh meat sausages can be grilled or fried while still frozen, and cook right through in a short time. Fat-sealed sausages can be reheated quickly and easily.

Sausages have another merit, in being adaptable meal-makers. They can be served hot or cold. They are equally useful as a main course or snack, for breakfast, packed meals or a picnic. They can also help to 'stretch' other meats in dishes such as casseroles. Since they can be cooked at any temperature from fairly high to low, they can be put in the oven with almost any other dish, to accompany it or to use the oven space to the full.

In all the ways above, home-made sausage meat and sausages can save you money and worry. The following are some well-known ways to use them:

Sausage meat

forcemeat balls
poultry and game stuffings
filling vegetables, *e.g.* stuffed
    tomatoes
galantines and brawns
pâtés and terrines
meat and game pies (to fill up
    spaces)
meat layers in meat and vegetable
    'bakes'
savoury pancake fillings

Small sausages

as cocktail 'nibbles'
to garnish soups
with baked beans
with red cabbage and apple
with sauerkraut
with meat, shellfish and rice (in paella
    and other pasta dishes)
in pizzas
as 'hot dogs'
as Sausage Hot Pot
in Bubble and Squeak

as Toad in the Hole
as a packed meal or picnic dish
as a kebab item
as a barbecue meat
as part of a mixed grill

Large sausages

as hors d'oeuvres (thinly sliced)
in salad (cubed or sliced)
as main cold meat dishes with salads and cheese (sliced)
with pasta or sauerkraut as main dish (cubed)
as layers in galantines, brawns, *etc* (sliced)
in toasted and other sandwiches (sliced)
boiled in chunks, with mashed potato (boiling rings)

The basic sausage meat recipe (page 54) and other sausage recipes in this book are suitable for most of the above dishes, but once you have decided to make sausages, you may want to try other types including Continental ones for slicing and salads. Some of their names are strange, and may be confusing, the following is a list of some common Continental sausages, what they contain and how they are used.

Some Continental Sausages

Bierwurst Typically German sausage, smoked and cooked, containing lightly minced pork, fine flavour, no garlic. Eaten cold, sliced. Try with beer and black bread.

Blutwurst Slightly smoked sausage containing very finely chopped pork and beef meat with diced pork back fat, meat juices and seasoning. Eaten cold or fried.

Cabanos Long, thin smoked and dried sausage, containing coarse-cut pork. Eaten cold or reheated for five minutes. Add in chunks to bean and pea soups 20 minutes before serving.

Chorizo Smoked and dried Spanish sausage of spiced fat and lean pork with liver or beef, well chopped and seasoned with Cayenne pepper, juniper, red peppers and tomato. Cooked with chickpeas (*garbanzos*) as a main dish, or used in stews.

Cervelat Smoked slicing sausage, containing specially dried pork meat, finely minced; mild seasoning. Eaten cold, preferably very thinly sliced. Good with rye bread and butter.

Extrawurst Finely milled pork, subtle seasoning. Eaten as a cold slicing sausage or diced in made-up dishes. Slices fried in butter will curl up to form a cup for savoury fillings.

Frankfurter The 'hot dog' choice, made of finely minced pork. Rather bland, but juicy, smoked and fully cooked. To heat through, bring water or white wine to the boil, turn off heat and leave for five minutes, then drain. Like all Continental sausages, it is 100 per cent meat.

Knackwurst Small dumpy sausage of finely minced pork meat, lightly smoked and seasoned. Eaten hot; often served with sauerkraut with garlic added to boost the flavour. Put in cold water, bring gently to boiling point. Leave in water off heat for 8–10 minutes.

Krakauer A Polish slicing sausage, cooked and smoked, containing coarse-cut ham and pork. Eaten cold. Good for open sandwiches or in 'closed' ones for the lunch box.

Leberwurst A smooth-textured sausage containing a pâté of finely milled calves' liver and pork meat. Pleasant starter to a meal with hot toast, butter and crisp radishes.

Liver Sausage Made of finely milled and coarse-cut liver and pork, in a natural casing. Eaten cold as a canapé spread, or hot, sliced, egg-and-crumbed and fried in a little butter.

Mettwurst A smoked ring sausage containing dried pork meat, coarsely chopped and seasoned. Eaten raw in sandwiches or boiled to accompany mashed potatoes or swedes.

Mortadella Smoked Italian sausage containing very finely milled pork, ham, and diced pork-back fat, flavoured with whole peppercorns, pistachio nuts and mild garlic. Eaten cold.

Pork Rings A larger smoked sausage containing coarse-cut pork meat and seasoning. Eaten hot—fried, sliced or boiled whole. Heat through like Knackwurst. A favourite with mashed potatoes and sauerkraut. When sliced, fries quicker than bacon for breakfast.

Salami Italian sausage. Many varieties. Most readily available here is Salame Milano, made of equal proportions of lean pork, beef and pork fat seasoned with pepper, garlic and white wine. It is smoked, eaten cold or hot in made-up dishes. Hungarian salami, made of pork, is highly regarded. Other salami come from Denmark and Germany, and are less highly flavoured than Italian salami.

Teewurst A mild sausage containing finely minced pork meat. Try it as a spread on thin toast.

Tongue Sausage Made of whole tongues of veal, pork and seasoning. Use it for Danish sandwiches.

Making Sausages in Skins

Most sausage meat mixtures are cased; that is, stuffed into a skin or casing. Some mixtures are too loose or soft to be cooked without a casing, and all sausages are stored more easily if cased. The word 'casing' strictly means the large or small intestine of the pig or other animal, or a long thin tube very like it made of cellulose. But other interior membranes, especially caul fat, are used as containers and may be referred to, informally, as casings too.

When you consider what kind of containers or casings to get for your sausages, think in terms of using pig's or sheep's casings, cellulose casings or caul fat. Other casings and containers, *e.g.* beef casings, make sausages too big for safe domestic handling. Even items such as a pig's bladder or sheep's stomach make quite a big 'sausage' which cannot be cut up before cooking, is unwieldy to store and is usually larger than an average family wants for one meal. You will find it much more convenient to make smaller sausages in conventional 'links' which are easily divided for use as individual portions, or to use caul fat for small packages of sausage meat.

Ways of obtaining various kinds of containers are suggested on page 46 above. But this is only the first part of your task. Even when processed for use, they must still be prepared for filling, and filled.

If a friendly butcher is supplying casings, the easiest way to get them filled is to take your sausage meat mixture to him and ask him to do the filling. If he cannot do it, ask his advice on how to process and prepare the casings he is supplying to you and what quantity you will need. This will depend on the size and type of casing he will supply and on the kind of mixture you have made.

If you have a large electric mixer with a mincing machine and a sausage filler attachment, you may prefer to fill your own sausage casings. First soak them as described on page 47 and rinse them in cold water. Then open each one out under a jet of water. A good way to do this is to turn on the cold water tap and with the water running push each length of casing in turn on to the end of the tap. Then assemble the sausage filler equipment according to the mixer manufacturer's instructions and fill the casings with your sausage meat mixture. Press out any air that accumulates in the casings and do not fill them too tightly or they will burst during cooking.

If you are using cellulose casings, see that your hands are dry and free from grease before stuffing them, or the sausage will have an uneven surface because of under-stuffing. Moisten either natural or synthetic casings after stuffing them, however, to make it easier to form them into 'links'. Do this by twisting the casings, or by tying off with soft string.

If the sausages are to be smoked (page 20), moisten the casings again just before smoking, to prevent them cracking due to uneven expansions during smoking. (Remember that home smoking, while it cooks and flavours the sausages, does not preserve them like the commerically smoked sausage. For that, you must take them to a local bacon factory for smoking. Home-made sausages, smoked at home, should either be eaten while still hot from the smoker, or be chilled quickly and then frozen for storage.)

Do not try to fill sausage skins at home without a proper filler. It is a frustrating task; besides, it is difficult to fill the casings evenly and without leaving air spaces, so they are likely to burst messily during cooking. It is wiser, and easier, to use caul fat to package the sausage meat if you have no filler equipment. Soak the stiff fat in lukewarm water and vinegar as described on page 47 above. When it is tender, smooth it out flat, and cut it into rectangles or squares the size you need. The sizes will, of course, depend on what size you want to make your sausages. But as a rough guide, average-sized flat *crépinettes* fit into 10–13 cm (4–5$\frac{1}{2}$ in.) squares, and so do the round faggots.

To fill the rectangle or square of caul fat, simply lay a lump of sausage meat in the centre and parcel it up so that the edges of the fat overlap or can be folded over the main parcel. Press them slightly to make them stick together, or bind them with a very little egg.

If you cannot get caul fat, you can make substitute 'skins' for baked sausages out of foil sheet cut into rectangles or squares; while, for a large boiling sausage, you can lay your meat mixture on a strip of well-floured cloth which you can then fold over the meat in the same shape as an old-fashioned roly-poly pudding. Tie soft string or tape round the cloth to keep it in place.

Sausages without Skins

Some experts say that caul fat is not a 'skin' and that sausage meat cooked in it is really made into skinless sausages. However, there are other, more obvious kinds of sausages without skins such as Spiced Farmer's Sausages (page 56). Even without skins, finely chopped or minced sausage meat mixtures containing flour or breadcrumbs will usually keep their shape if baked, grilled or gently fried, especially if coated with egg and breadcrumbs or flour before cooking.

Alternatively, you can make a sausage meat mixture which contains egg as a 'binder'. Remember, though, that it will not be suitable for smoking, freezing or long storage. Sausage meat containing egg should be eaten as soon as it is made, either as patties or as a stuffing.

Sausage Meat

Whether you use skins or not is less important than the quality of the sausage meat filling. No matter whether you make a standard pork sausage meat using all pork or make a meat mixture, use pork from a heavy hog rather than from a young porker if you can. It will be fattier and more mature in flavour. Use a recipe in which fat provides at least a quarter of the weight of the meat, *e.g.* 100 g (4 oz.) fat to 300 g (12 oz.) lean pork; sausage meat is an excellent way to use up unwanted fat in a bulk order. Season the mixture really well, especially if you are making sausages for immediate eating or a Continental-type sausage meat; vary the usual seasonings with more exotic ones sometimes. Nuts (chestnuts, pine nuts), vegetables (spinach, peppers, celery), fresh herbs (parsley, sage, thyme, wild garlic, rosemary, marigold petals) all add interesting flavour when mixed with the more usual seasonings. So can spices such as ground ginger, crushed clove, crushed scorched coriander and paprika. Use one or more to replace the standard allspice flavouring.

Finally, choose a recipe with little or no cereal in it. Commercial sausages in Britain contain as much as one-third of their weight in flour, rusk or other cereal. One of the major advantages of making your own sausages is that you can give your family a protein-packed, all-meat product for their meals cheaply. You will be wasting a good part of your efforts if you only repeat what the commercial manufacturers do.

Begin with the basic sausage meat recipe below if you are making sausage meat for the first time. When you have tested the various ways in which you can use your product, try other recipes. Look, for instance, at books on foreign and regional cookery. You will find in many of them recipes for interesting Continental-type sausages for which there is no space here.

**Stuffing a sausage skin.
The first stage**

*South African Livestock and
Meat Industries Central Board*

Basic Sausage Meat

480 g (1 lb.) lean fresh raw pork (see note)
200 g (8 oz.) firm pork fat without gristle
1 × 5 ml spoon (1 teaspoon) salt (see note)
1 × 2.5 ml spoon ($\frac{1}{2}$ teaspoon) ground allspice
Freshly ground black pepper to taste
Pinch of crushed dried marjoram or pennyroyal
25 g (1 oz.) stale white breadcrumbs (optional)

Mince the meat and fat twice. Mix thoroughly and season to taste with the spices and herbs. Mix in the breadcrumbs if used, adjust the seasoning and use the sausage meat as desired.
Note: If you wish, use lightly salted pork and reduce or omit the salt.

Basic Sausage Meat with Egg

(for skinless sausages, stuffings, *etc*)

Use the recipe for Basic Sausage Meat above, adding one lightly beaten egg when mixing the meats. Double the amount of salt and allspice.
Note: Try substituting one of the alternative spices suggested on page 52 above for the allspice, especially for a stuffing.

Sausage Patties or Crépinettes

300 g (12 oz.) Basic Sausage Meat with Egg
100 g minced beef, veal or chicken
1 × 5 ml spoon (1 teaspoon) finely chopped parsley or other herb suggested on page 52
Caul fat as required

Mix the sausage meat thoroughly with the other meat and the herb desired. Form into lumps and flatten them into oval patties. The size will depend on what they will be used for. 50 g (2 oz.) patties are usually big enough for a breakfast dish, while larger ones are usually wanted for a snack or main course dish.

Cut rectangles of softened caul fat which will encase the patties. Lay a patty in the centre of a piece of fat and fold the long sides of the fat over the patty so that it overlaps. Fold the shorter ends of the fat over to form a small parcel. Press to seal.

Either brush with beaten egg, coat with breadcrumbs and fry or grill gently, turning once; or refrigerate for 24-48 hours, then simmer in stock for about 40 minutes, pricking them with a skewer if they swell up.
Note: The cooking time will depend on the thickness of the patties. Test by slitting the centre of one patty with a knife to make sure there is no raw meat in the centre.

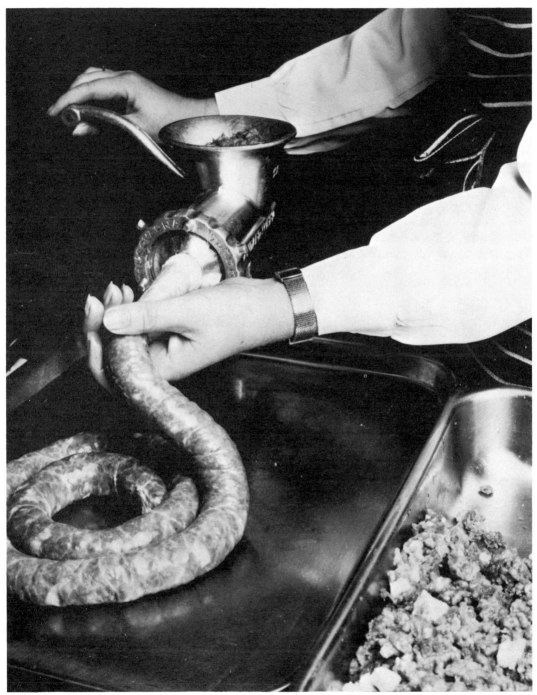

**Stuffing a sausage skin.
The second stage**

*South African Livestock and
Meat Industries Central Board*

Spiced Farmer's Sausage

Makes 6 × 75 g (3 oz.) sausages

100 g (4 oz.) minced fresh or lightly salted pork
100 g (4 oz.) minced stewing veal
100 g (4 oz.) shredded suet
Breadcrumbs from 3 slices dark brown wholewheat bread
1 × 5 ml spoon (1 teaspoon) grated lemon peel
1 × 2.5 ml spoon ($\frac{1}{2}$ teaspoon) ground dried sage
Good pinch each of ground dried marjoram, grated nutmeg and freshly ground pepper
1 × 2.5 ml spoon ($\frac{1}{2}$ teaspoon) salt if required
1 egg yolk
1 × 10 ml spoon (2 teaspoons) orange juice
Margarine or oil for greasing

Mix all the ingredients thoroughly, using an electric or rotary beater if possible. Form into six sausage shapes. Fold a large rectangle of foil lengthways to form a long, double-thick strip. Grease it well. Cut it into six rectangles. Wrap each sausage in a rectangle of foil, twist the ends of the foil to seal it securely, and lay the sausages in a small roasting or dripping tin. Bake for 30 minutes at 190°C, 375°F, Gas 5. Unwrap and serve hot at once, or leave to cool, and slice the sausages for a packed meal.
Note: If you wish, you can chill and store the sausages, then fry or grill them lightly when required.

Black Bear Sausage

12–16 sausages or patties

1 finely-chopped small onion
1 clove garlic, crushed
1 × 15 ml spoon (1 tablespoon) salt
1 × 5 ml spoon (1 teaspoon) saltpetre
1 × 5 ml spoon (1 teaspoon) coarsely ground black pepper
$\frac{1}{2}$ × 2.5 ml spoon ($\frac{1}{4}$ teaspoon) white pepper
1 × 5 ml spoon (1 teaspoon) crushed juniper berries
1 × 15 ml spoon (1 tablespoon) light soft brown sugar
250 ml ($\frac{1}{2}$ pint) water
700–750 g (1$\frac{1}{2}$ lb.) minced beef
300 g ($\frac{3}{4}$ lb.) minced pork
100 g (4 oz.) minced veal
1 × 15 ml spoon (1 tablespoon) paprika

Place the onion, garlic, salt, saltpetre, peppers, juniper and sugar in a saucepan. Pour on the water and leave for five to six hours, stirring occasionally until the salt, saltpetre and sugar have dissolved.
 Warm the liquid gently just to boiling point. Strain through a fine sieve and cool to room temperature.
 Mince the meats a second time if at all coarse. Mix together thoroughly, with the spicy liquid and paprika; this is best done with wet hands. Adjust the seasoning if desired, then form the mixture into 75 g (3 oz.) long rolls like sausages, or into round patties.
 Either bake at 180°C, 350°F, Gas 4 for 35–45 minutes in an ungreased baking or dripping tin, or smoke in a home smoking kit for about $\frac{1}{2}$ hour. The baked sausages give a lot of

flavoured fat. The smoked ones stay fattier and moister in texture. Serve hot or cold.

Note: If you use half the mixture for baked sausages and half for smoked ones, you will have two products completely different in flavour and texture. Served cold, sliced, in separate piles on the same dish, they make a complete first course or buffet party dish. Both kinds keep well for several days if refrigerated. To reheat, wrap in foil, and bake until just heated through.

Faggots

Four helpings.

480 g (1 lb.) pig's liver or pig's fry (lites, liver, heart, melts) or choice of these
3 onions
75 g (3 oz.) breadcrumbs or boiled potato
25 g seasoning, including salt and white pepper, crushed dried sage and sweet basil and a pinch of ground ginger
Flour as required
Lard as required

Simmer the meat and onions in salted water until cooked. Drain. Use a little of the cooking liquid to moisten the breadcrumbs or potato. Mince the meat, onion and crumbs or potato together, then pound or process in an electric blender to a smooth paste. Season with the ingredients above, or to taste. Divide the mixture into eight round balls. Roll them in flour.

Grease a baking tin or dripping tin which will just hold the balls side by side. Place the balls in the tin. Cover them loosely with a piece of foil. Bake at 180°C, 350°F, Gas 4 for 25 minutes. Remove the foil and bake for a further seven to 10 minutes. Divide the balls with a knife if they have stuck together, and serve hot with a gravy made from the cooking liquid and with tomato sauce.

French Faggots

Four to six helpings.

480 g pig's liver and fry (lites, heart, melts, *etc*) cleaned and without gristle
150–200 g (6–8 oz.) Basic Sausage Meat
2 cloves garlic, crushed
Salt and freshly ground black pepper to taste
Spices to taste (see page 52)
1 × 5 ml spoon (1 teaspoon) chopped fresh parsley
Caul fat as required
Lard for greasing

Mince the liver and fry and mix thoroughly with the sausage meat and seasonings. Form into the shape of small golf-balls. Cut squares of caul fat which will encase the balls and wrap each ball securely in a piece of fat. Grease a small roasting tin or dripping tin with lard and lay the faggots in it side by side. Melt a little extra lard and pour it over them. Bake at 180°C, 350°F, Gas 4 for about 40 minutes. Raise the heat for the last five minutes to brown the tops of the balls if you wish. Serve hot; or allow to get quite cold, wrap each ball in foil and use as a packed meal or for a picnic.

South African Sausage or Boerewors

3 kg (about 6½ lb.) stewing beef
2 kg (about 4¼ lb.) pork
30 g (1½ oz.) scorched crushed coriander (see recipe)
25–50 g (1–2 oz.) fine salt
1 × 5 ml spoon (1 teaspoon) pepper
250 g (10 oz.) pork fat
150 ml (6 fl. oz.) vinegar
100 g (4 oz.) sausage casings

Cut the beef and pork into ½ cm (¼ in.) dice. Scorch the coriander by heating the whole berries gently in a dry frying pan, stirring constantly, until lightly browned; then grind to coarse grains in an electric blender or coffee mill. Sieve to remove husks and very large grains.

    Mix the diced meats, coriander, salt and pepper lightly. Mince the mixture. Using a large-holed screen, mince the pork fat separately. Mix together, then mix in the vinegar as lightly as possible. The texture of the sausages should be loose and coarse, not tight-packed. Fill into the casings as described earlier.

Mutton Boerewors

500 g (1 lb.) beef
500 g (1 lb.) pork
1 kg (2¼ lb.) lamb
1 × 10 ml spoon (1 dessertspoon) fine salt or to taste
Pinch of ground scorched coriander (see South African Sausage)
Good pinch of pepper
Pinch of ground cloves
250 g (10 oz.) firm pork fat
25 ml wine vinegar
250 ml sweet red wine and brandy, mixed
75 g (3 oz.) sausage casings

For the quality of meat to use, see page 52. Make the sausage in the same way as the South African Sausage using wine and brandy with the vinegar.

Potted Meats

'Potting' meats became popular in Tudor times, along with the fashion of sealing solid meats in fat (page 29). It was a way of making short-term preserves out of the bits and pieces of a carcase, especially offals.

These bits and pieces had good food value, but they tainted quickly. So they had to be eaten or preserved at once. They were not attractive or solid enough to process for long-term keeping, like joints, so they were pounded and cooked with fat, to make them last just for a few days or weeks. They would be useful for eating while the larger pieces of meat were being cured.

True potted meats are still one of the most useful and pleasant meat preserves to make at home. One can have fun collecting attractive small pots in various colours for one's 'pottery'. Very cheap pieces of meat can be used and potted meats are ideal for using up leftovers. Most people like them since they are easy to eat and make good snack and packed meals; they keep for a week or more in a refrigerator without coming to harm, so are a good kitchen standby.

They also make an intriguing party buffet. Make an assortment from leftovers during the previous week. Label them, supply bread, butter and knives and tell guests to choose their own. They will be interested by the labels and flavours and you will have no serving problems. (Small whisky glasses make good small pots for a party.)

Potted meats are really just another form of meat sealed in fat (page 29). So they are very quick and easy to make. They are all made in much the same way. The meat is cleaned of gristle and bone. It is cooked, then minced or pounded with fat and flavouring until it is a paste. Then it is pressed well down into small, clean pots, which are sealed twice with clarified fat.

Unlike pâtés, which are described in the next section, these meat pastes do not all, by custom, contain pork, ham or bacon. Because of this, we have developed a tradition of mixing and sealing the meat with clarified butter rather than lard or pork fat.

The butter is clarified to cleanse it of any foreign matter and excess salt. Even a minute speck of foreign matter can hold bacteria and may transfer them to the potted meat instead of protecting it. As for salt, potted meats are best kept in a refrigerator and so should not be kept under heavily salted sealing fat. Besides, even if one breaks the butter seal without meaning to eat it, bits of butter always get mixed up with the potted meat itself; since most potted meats are quite strongly flavoured already, they may be spoiled by the extra seasoning. Pure fat also makes a smoother, clean-looking coating. So it is safer and wiser to use clarified butter or other fat even if it seems wasteful.

All fats are clarified in the same way. Here is a standard method of doing it, followed by some traditional recipes for British potted meats.

To Clarify Butter or other Fats

Chop the fat into small pieces, if hard. Put enough water in a heavy-bottomed saucepan to cover the bottom. Add the fat and melt it very gently over a low heat. When it is fully melted, remove the pan from the heat and allow it to stand undisturbed for several minutes until any foreign matter sinks to the bottom. Salt butter will have a clear golden lay on top with white sediment underneath. Dripping may be amber, with darker meat juice beneath.

If you have a steady hand, lay a double thickness of butter-muslin or old sheet in a sieve placed over a basin and pour the clear top layer of fat through it into the basin, leaving the sediment behind. Otherwise let the fat get cold and firm, then ease it out of the saucepan with a knife and scrape the bottom free of sediment.

Prepare as much clarified fat as you can at one time, at least 500 g (1 lb.) if possible. The larger the quantity, the easier it is to make and it keeps excellently in the refrigerator. Store it in a covered basin and take out as much as you need for use with a clean spoon dipped in boiling water.

Note: Dripping or lard can be placed in a clean dripping or baking tin instead of a saucepan and be clarified by heating in a low oven when you happen to be using it.

Potted Beef 1

A good way to use up the tough odds and ends of meat one sometimes has to accept with a large joint or with part of a carcase bought for freezing.

Four helpings.

480 g (1 lb.) shin of beef or similar meat.
2 bay leaves.
2 cloves.
Pinch of ground mace.
2 × 15 ml spoons (2 tablespoons) water.
Salt and pepper to taste.
About 75 g (3 oz.) butter.
Clarified butter to cover.

Cut the meat into small pieces, removing any fat and gristle. Place it in an oven-proof dish with the bay leaves, cloves and water. Cover closely with buttered paper and kitchen foil. Bake at 150°C, 300°F, Gas 1–2 for 3–3½ hours until the meat is very tender. Remove the bay leaves and cloves. Mince the meat twice then pound it well with any remaining meat juices and the butter, to make it a smooth paste. Add salt and pepper to taste. Press the meat into pots, tapping on the table two or three times while filling, to knock out any air bubbles. Leave a good 1 cm (½ in.) head-space. Cover with melted clarified butter and allow it to cool and firm up; then re-seal with a second thin coating of clarified butter to seal any hair cracks due to shrinkage.

Potted Beef 2

Pieces of cold roast beef
Quarter the weight of the beef in butter
1 × 5 ml spoon (1 teaspoon) anchovy sauce for each 25 g (1 oz.) butter
½ × 2.5 ml spoon (¼ teaspoon) each pepper and ground mace for each 25 g (1 oz.) butter
Pinch each of grated nutmeg, Cayenne pepper and salt for each 25 g (1 oz.) butter
1 small lump of sugar for each 25 g (1 oz.) butter
Clarified butter to cover

Clean the beef of excess fat and any skin and gristle. Mince it and weigh it, then measure the butter. Heat the butter and seasonings gently in a saucepan until melted. While it heats, pound the meat or process it in an electric blender to a smooth paste.

Mix the meat into the melted butter until well blended. Let the meat heat through thoroughly. Then press into pots, tapping on the table top two or three times while filling to knock out any air bubbles. Leave a good 1 cm (½ in.) head-space. Cover with melted clarified butter. Allow to firm and cool; then re-seal with a second coat of melted clarified butter.

Potted Lamb

Make like the beef above, but add a little crushed dried thyme or rosemary.

Potted Pork

480 g (1 lb.) lean cold cooked pork without crackling
75 g (3 oz.) butter.
Pinch of grated onion.
Good pinch of dried sage.
Salt to taste.
Clarified lard to cover.

Mince the pork, then pound it with the butter and seasonings, or process in an electric blender. When smooth and pasty, taste and adjust the seasoning. Press into small pots, tapping on the table top occasionally while filling to knock out any air bubbles. Leave a good 1 cm (½ in.) head-space. Cover with melted clarified lard.

Potted Ham or Bacon

480 g (1 lb.) cold lean boiled ham or bacon
100 g (4 oz.) unsalted butter
Ground mace to taste
Clarified bacon fat to cover

Pound the ham or bacon with the butter, or process in an electric blender. Season to taste. When the mixture is a fine paste, press it into small pots, tapping on the table two or three times while potting, to knock out any air bubbles. Leave a good 1 cm (½ in.) head-space. Cover with melted clarified fat. When cold and firm, re-seal with a second thin layer of fat.

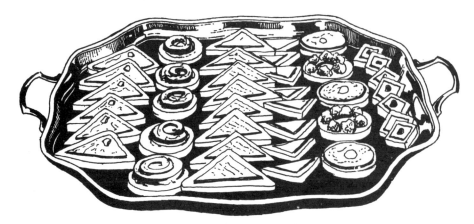

2. Potted Meats used on Canapés

Potted Veal

500 g (1 lb.) cold roast or braised veal
100 g (4 oz.) lean ham or bacon
2 × 10 ml spoons (2 dessertspoons) strong veal stock or gravy
Pinch each of salt, pepper and ground mace
25 g (1 oz.) butter
Clarified butter to cover

Remove any skin, gristle and bones from the meat. Place the veal and the ham or bacon in a small saucepan, barely cover them with water, add a little salt, cover and simmer until well heated through. Drain, then mince both meats. Pound them or process in an electric blender with the pepper and mace, butter and stock or gravy. When pounded to a smooth paste, taste and adjust the seasoning. Press into small pots, tapping on the table two or three times while filling to knock out any air bubbles. Leave a good 1 cm ($\frac{1}{2}$ in.) head-space. Cover with melted clarified butter. When firm and cold, re-seal with a second thin coat of clarified butter.

Potted Tongue—a recipe of 1845!

Boil tender an unsmoked tongue of good flavour and the following day cut from it the quantity desired for potting, or take for this purpose the remains of one which has been served at table. Trim off the skin and rind, weigh the meat, mince it very small, then pound it as fine as possible with 100 g (4 oz.) butter to each 480 g (1 lb.) of tongue, 1 × 5 ml spoon (1 small teaspoon) ground mace, 1 × 2.5 ml spoon ($\frac{1}{2}$ teaspoon) each of nutmeg and cloves (ground) and a tolerably high seasoning (a good sprinkling) of Cayenne. After the spices are well beaten with the meat, taste it and add more if required. A few grams (ounces) of any well-roasted meat mixed with the tongue will give it firmness in which it is apt to be deficient. The breasts of turkeys, fowls, partridges or pheasants may be used for this purpose with good effect.

(Pot and seal with melted clarified butter exactly like the recipes above.)
Note: I have given this recipe exactly as the old author wrote it except for adding modern measures.

1. Antique pestle and mortar for pounding meats

Potted Pheasant

1 roasted pheasant or the remains of a brace
½ bay leaf
100 ml (4 fl. oz.) sherry
2 shallots
1 sprig of thyme
450 ml (1 pt.) good game stock
100 g (4 oz.) butter
Salt and Cayenne pepper to taste
Clarified butter to cover

Remove the bones and any skin and sinew from the meat. Place it in a saucepan with the bay leaf, sherry, shallots, thyme and stock. Simmer until the liquid is a glaze. Strain and remove the bay leaf and thyme; reserve the glaze. Pound the solid ingredients or process in an electric blender to a smooth paste. Blend in the glaze and butter thoroughly and season to taste. Press into small pots, tapping on the table while potting to knock out any air bubbles. Leave a good 1 cm (½ in.) head-space. Cover with melted clarified butter. When cold and firm, re-seal with a second thin coat of clarified butter.

Potted Rabbit

1 rabbit
Salt as required
50 g (2 oz.) unsalted butter
1 × 5 ml spoon (1 teaspoon) caster sugar
1 small onion
8–12 cloves
8–12 allspice berries tied in a rag
Grated nutmeg or ground mace to taste
150 g (6 ozj) slightly salted butter
1 × 10 ml spoon (1 dessertspoon) Worcestershire sauce
Clarified butter to cover

Joint the rabbit and soak it in salted water for 2–3 hours. Drain and pat dry. Place it in a heavy-bottomed casserole with the onion (stuck with the cloves) and the spices. Cover securely with a lid or foil and cook at 150°C, 300°F, Gas 1–2 for 2½ hours. Allow to cool. Remove the meat from the bones and mince it twice. Mix thoroughly with the juices from the casserole, the 150 g (6 oz.) butter and the Worcestershire sauce. Press into small pots, tapping on the table top two or three times while filling to knock out any air bubbles. Leave a good 1 cm (½ in.) head-space. Cover with melted clarified butter. When cold and firm, re-seal with a second thin coat of clarified butter.

Pâtés and Terrines

Pâtés and terrines, unlike potted meats, are processed before cooking, so they can be served in the dishes they are cooked in. Both are minced or pounded meat mixtures, sometimes with layers of sliced solid meat between the softer minced layers. All the same kinds of meat are used as in making potted meats, and for the same reasons.

Pâtés and terrines vary in their contents even more than potted meats. The basic meat mixture of either can be almost anything you like. So can the spices, herbs, alcohol and other ingredients added to it. Raw or cooked meat can be used, so pâtés and terrines are an ideal way to use up bits and pieces of meat in a bulk order, or leftovers. Some pâtés in pastry can be served hot.

The word 'pâté' comes from the same word as *paste, pastry* and *pasta*. It can mean a soft filling or a pastry dough. In fact, soft meat pâtés were originally cooked in a pastry case. In Tudor times, a solid hot water crust pastry was used because all pies and pasties then were 'raised pies'; our Pork Pie has survived in that shape. Later, short crust or flaky pastry came into fashion for encasing pâtés. But it only stayed in use for very luxurious pâtés, which the French call dishes *en croûte*. A jacket of sliced pork fat was found cheaper and better for protecting everyday pâtés from dryness caused by the cooking heat.

Today, we often use overlapping slices of fatty bacon instead of pure fat; they are used to line the dish before putting in the main meat mixture. Slices of back pork fat may still be put on top of the pâté and are removed after it is cooked. But as a rule, a pâté is turned out of the cooking dish on to a serving platter after being cooked and cooled and the bacon strips lining the dish are then kept as decorative covering. (See the picture on page 69.)

A terrine has, traditionally, always been cooked in an earthenware dish, not in a pastry crust. ('Terrine' is the French name for the earthenware dish.) In the past, the meat mixture was always served in this cooking dish and it is still served this way quite often. But, today, it may also be turned out like a pâté. In fact, there is now really no difference between a pâté and a terrine.

Like potted meats, both pâtés and terrines make good nourishing meat dishes at little cost. It is not practical to make small ones, so the smaller jars of potted meats may be more useful for one or two people. But pâtés and terrines are of more value to a family or to anyone who gets bulk supplies of meat.

All ordinary pâtés and terrines are cooked in the same way. First, grease a suitable dish or tin (see below) or line it with sliced pork fat, pickled pork or bacon. Then fill it with your chosen meat mixture or with alternate layers of mixture and sliced meat. The top is usually covered with more pork fat or bacon and with two or three bay leaves. Seal the dish securely and place it in a pan of very hot water, which reaches about half-way up its sides. Then bake it; the time and heat depend on the kind of meat mixture you use. Add more water to the pan during baking if required.

The mixture is ready when it begins to shrink from the sides of the dish or tin and the liquid fat which has risen to the top is clear. Remove the dish from the oven and from the pan of water. Uncover it and take off the bay leaves and the pork fat or bacon on top. Re-cover the meat mixture with grease-proof paper and place a weight on top of it which presses it down evenly all over. To make a weight, cut a piece of stiff cardboard which fits inside the top of the dish or tin. Place this on the greaseproof paper and put a heavy weight such as a flat iron or a large stone in the centre of the dish. Leave the weighted dish in a cool place for 24–48 hours until the flavours have matured. No pâté or terrine should be eaten straightaway after being made.

The matured pâté or terrine can then be turned out for serving, or just to remove the pork fat or bacon lining the dish. This is usually done by running a sharp knife round the sides of the dish to loosen the mixture from it; a platter is placed upside down on the dish, which is then turned over and jerked so that the pâté mixture drops on to the platter. Sometimes the cooking dish may have to be dipped in boiling water first. When turned out, the fat pork slices or bacon rashers which lined the dish can be stripped off if you wish. The pâté or terrine can then be sliced and served, or it can be returned to its own dish for serving.

If you prefer, you can serve the pâté or terrine from its own dish without turning it out first. In this case, remove the top fat or bacon only and replace the bay leaves on the meat mixture to garnish it before covering it with greaseproof paper. Remove the strips of pork fat or bacon when cutting slices for serving.

A pâté *'en croûte'* is cooked and cooled differently. The method will depend on whether you want to make an old-style raised pie in a hot water crust, or a pâté shaped like a deep plate pie, pasty or meat roll, using short crust or flaky pastry. In all these cases, follow a standard recipe for the type you want to make. These pâtés are not included in the recipes below because they are not preserved meats in the same way as ordinary pâtés. Pastry can soon go stale and it does not protect and seal the meat mixture as pork fat or bacon does. Pâtés *en croûte* should be treated as pies and eaten within 36 hours of being baked.

If you want to make old-style pâtés and terrines to serve in the dish, it is worthwhile investing in the proper 'cook-and-serve' dishes. They are admirably suited to their purpose and look attractive on the table. Pâtés and terrines for turning out can quite well be cooked in a loaf or cake tin or in a foil pan and they will take less space in the freezer if you have to store the goods for long. Do not freeze pâtés unless you have to. They tend to go damp and taste 'dead'. Most pâtés and terrines will keep well in a refrigerator for a week or longer; the exact time will depend on the ingredients used.

Although pâté and terrine mixtures vary widely, some are grouped together because they have the same contents or texture. Their names are useful to know when deciding what pâté to make. A *pâté maison*, for instance, is a pâté made to a particular cook's own recipe, so its ingredients may differ widely, but it nearly always has the same slightly coarse, loose texture since its name described pâtés once made at home without modern aids like a mincer. A *pâté de campagne* (which means 'pâté of the district or region') is even more likely to be a crumbly, coarse-textured pâté.

Pâté mixtures can be set in aspic jelly after being cooked. This makes them a class of pâtés on their own. They are good; but the jelly does not help to preserve the meat as well as fat does, so (like pâtés *en croûte*) they are not included in the recipes below.

These recipes describe the more usual kinds of pâté mixture, minced, layered with solid meat and so on. Use them as models, changing the ingredients to any meat or seasoning which suits you. The ingredients in the recipes are standard ones, but many other combinations are used just as often.

Meat Pâté 1

150–200 g (6–8 oz.) back bacon rashers without rinds
100 g (4 oz.) calf's or lamb's liver, chopped
100 g (4 oz.) lean raw pork, chopped
100 g (4 oz.) pork sausage meat
25 g (1 oz.) soft white breadcrumbs
1 × 15 ml spoon (1 tablespoon) milk
1 small onion, finely chopped
1 small egg, beaten
50–75 ml (2–3 fl. oz.) brandy
Salt and pepper to taste
Pinch of grated nutmeg
200 g (8 oz.) cold cooked veal, chicken or rabbit, sliced

Line a 1 kg (2 lb.) loaf tin with the bacon rashers. Mix together the liver, pork, sausage meat, breadcrumbs, milk, onion and egg. Add enough brandy and seasoning to make a soft, well-flavoured mixture. Place an even layer of the sliced meat in the tin and cover it with an even layer of the sausage meat mixture. Repeat the layers until the tin is full, ending with a sliced meat layer. Seal the tin securely, with foil and stand it in a pan of very hot water. Bake at 170°C, 325°F, Gas 3 for $2\frac{1}{2}$–3 hours. Remove from the oven, place a weight on top as described on page 65 and leave in a cool place for at least 24 hours. Turn out as described on page 65.

Meat Pâté 2
(using cooked meat)

200 g (8 oz.) cooked beef or lamb without skin or gristle
150 g (6 oz.) beef or pork sausage meat
2 slices stale white bread without crusts.
Milk as required (see recipe)
2 small onions, finely chopped
3 × 15 ml spoons (3 tablespoons) finely chopped fresh parsley
1 × 10 ml spoon (1 dessertspoon) crushed dried chervil
2 egg yolks
Salt and ground black pepper to taste
Beef or lamb dripping for greasing
4 gherkins, finely chopped
2 hard-boiled eggs

Mince the meat twice and mix well with the sausage meat. Moisten the bread with milk and squeeze it dry. Add the squeezed bread to the meats, with the onions, herbs and egg yolks. Mix all together thoroughly and season well.

Grease a loaf tin well with dripping. Turn in one-third of the meat mixture and press down in an even layer. Sprinkle with a layer of chopped gherkin. Cover with a thin layer of meat mixture, filling about half the tin. Lay the whole hard-boiled eggs on this, end to end and about 2.5 cm (1 in.) apart. Cover them with meat mixture and level the surface all over. Sprinkle with a second layer of chopped gherkin, then fill the tin with the remaining meat mixture. Press it down evenly all over. Cover with foil and seal securely. Stand the tin in a pan of hot water and bake at 180°C, 350°F, Gas 4 for about one hour. Remove from the oven and from the pan of water. Place a weight on top as described on page 65 and leave in a cool place for 24-48 hours before use. To use, remove the weight and foil covering. Run a sharp knife round the inside of the tin to loosen the pâté from its sides, turn out as described on page 65 and sprinkle with any remaining gherkin before serving.

Liver Pâté

500 g (1 lb.) calf's, pig's or poultry liver
100 g (4 oz.) streaky bacon without rinds
1 small onion
50 g (2 oz.) butter
2 hard-boiled eggs, chopped
Salt and pepper to taste
Pinch of dried mixed herbs
1 × 10 ml spoon (1 dessertspoon) brandy
2 × 10 ml spoons (2 dessertspoons) cream
Butter for greasing

Chop the liver and the bacon finely. Skin the onion and chop it. Melt the butter and fry the liver, bacon and onion gently for four to six minutes, turning them over to brown the liver on all sides. Remove from the heat and add the hard-boiled eggs to the mixture in the pan. Pound the mixture, mince it twice or process in an electric blender to obtain a smooth paste. Add the seasoning, herbs, eggs, brandy and cream. Grease a loaf tin with butter. Put in the mixture and seal securely with foil. Stand the tin in a pan of very hot water and bake at 180°C, 350°F, Gas 4 for 40-45 minutes. Remove from the oven and from the pan of water. Place a light weight on top as described on page 65 and leave in a cool place or in a refrigerator for 36 hours before use. To use, remove the foil, run a sharp knife round the sides of the tin to loosen the pâté from it and turn out on to a serving dish as described on page 65.

Bacon and Liver Pâté

200 g (8 oz.) back bacon rashers without rinds
Bacon fat for greasing
200 g (8 oz.) fat piece forehock of bacon
50 g (2 oz.) butter
200 g (8 oz.) pig's liver
1 clove garlic, chopped
1 large onion, chopped
Salt and pepper to taste

For the sauce:
250 ml ($\frac{1}{2}$ pint) milk
2 blades mace
1 bay leaf
2–4 peppercorns
25 g (1 oz.) butter
25 g (1 oz.) flour

Stretch the bacon rashers on a board with the back of a knife. Grease a straight-sided ovenproof dish with bacon fat. Lay the rashers on the bottom and round the sides of the dish to line it. Cut any rind off the forehock. Heat the butter in a frying pan and fry the forehock, liver, garlic and onion gently for 10 minutes, turning the mixture over from time to time. Mince finely or process in an electric blender.

To make the sauce, put the milk in a saucepan with the mace, bay leaf and peppercorns. Bring slowly to the boil, remove from the heat and allow to stand for ten minutes. Melt the 25 g (1 oz.) butter in a clean pan, add the flour and stir well to blend. Cook for one minute. Remove from the heat and strain in the milk gradually, stirring to prevent lumps forming. Return to the heat and bring gently to the boil, stirring continuously, until the sauce thickens and bubbles. Add the sauce to the liver mixture and blend in thoroughly. Season to taste.

Turn the mixture into the bacon-lined dish. Cover closely with foil and with a lid if possible. Stand the dish in a pan of hot water and bake for one hour at 180°C, 350°F, Gas 4. Remove from the oven and from the pan of water. Place a weight on top as described on page 65 and leave in a cool place for 36–48 hours before use. To use, run a sharp knife round the sides of dish and turn out the pâté with the bacon rashers covering it, as described on page 65.

Bacon Terrine

6 rashers streaky bacon without rinds
200 g (8 oz.) pig's liver
1 onion, skinned
1 clove garlic, peeled
200 g (8 oz.) pork sausage meat
200 g (8 oz.) pork fat or pieces, minced
Salt and pepper to taste
2 hard-boiled eggs, chopped
1 × 15 ml spoon (1 tablespoon) chopped fresh herbs
200 g (8 oz.) veal or game (*e.g.* rabbit), minced coarsely
3 bay leaves

Bacon and Liver Pâté *British Bacon Curers' Federation*

Stretch the bacon rashers with the blade of a knife and line a straight-sided terrine with them. Mince the liver with the onion and garlic and mix these thoroughly with the sausage meat and pork. Season well and add the chopped eggs and herbs. Place a layer of this mixture in the bottom of the terrine, spreading it evenly. Cover the mixture with an even layer of the veal or game meat. Repeat these layers until the dish is full, ending with a layer of sausage meat mixture. Place the bay leaves on top in a decorative pattern.

Cover the terrine securely with foil and with a lid if possible. Stand the terrine in a pan of very hot water and bake at 190°C, 375°F, Gas 5 for 1–1½ hours. Remove from the oven and from the pan of water. Place on top a weight of about 1 kg (2 lb.). Leave in a cool place for 36 hours before use. To use, remove the foil carefully and serve in the terrine.

Note: If desired, this terrine can be turned out and used as pâté.

Chicken Terrine

1 × 1½ kg (3 lb.) chicken
150 ml (6 fl. oz.) white wine
Pinch of Cayenne pepper
Salt and pepper to taste
200 g (8 oz.) fresh belly of pork without bone
200 g (8 oz.) pickled belly of pork without bone
3 shallots
½ clove garlic
2 eggs
15 g (½ oz.) mushroom peelings
Pinch of mixed spice
500 g (1 lb.) hard pork fat, sliced

Skin the chicken. Take off the legs and wings and slice their flesh into thick matchsticks. Soak them in the wine with the Cayenne pepper, salt and pepper for about two hours.

Remove all the rest of the chicken meat from the carcase and mince it coarsely with the fresh and pickled pork, the shallots and garlic. Add the eggs and mushroom peelings, then mix in the wine the chicken flesh has been soaking in. Season with salt, pepper and spice. Mix well, taste and add extra seasoning if required.

Line a terrine with most of the pork fat, keeping aside enough to cover the top of the pâté. Put in a layer of the chicken matchsticks, then a layer of the minced meat. Repeat these layers until the dish is full, ending with a layer of minced meat. Cover with the remaining pork fat. Seal the terrine securely with foil and with a lid if possible. Stand the terrine in a dish of very hot water and bake at 180°C, 350°F, Gas 4 for about 1½ hours. Remove from the oven and from the pan of water. Place a weight on top as described on page 65 and leave in a cool place for at least 24 hours. To use, remove the foil and the top pork fat carefully. Serve in the dish, removing the pork fat which lines the terrine when slicing the pâté.

Game Pâté

480 g (1 lb.) stewing venison or other game meat without bones
1 rasher streaky bacon without rind
25 g (1 oz.) unsalted butter
50 g (2 oz.) button mushrooms, sliced
1 × 1 cm ($\frac{1}{2}$ in.) slice bread without crusts
Milk as required (see recipe)
50 g (2 oz.) slightly salted butter, softened
1 egg
90 ml ($3\frac{1}{2}$ fl. oz.) medium-sweet sherry
Salt and freshly ground black pepper to taste
Good pinch of grated orange peel
6 back bacon rashers without rinds

Mince or shred the game meat and streaky bacon finely. Melt the unsalted butter and sauté the mushrooms gently until tender. Moisten the bread with a little milk, then squeeze dry. Mix or pound together the meats, mushrooms and the bread, to the consistency you want, either a smooth paste or a coarser one with fragments of mushroom. Mix in thoroughly the salted butter, egg, sherry and seasonings. Stretch the back bacon rashers with a knife blade and line a loaf tin with them. Fill the tin with the meat mixture. Seal it securely with foil and place it in a pan of very hot water. Bake at 170°C, 325°F, Gas 3 for two hours or until the meat 'tests done' (page 65). Remove the tin from the oven and from the pan of water. Place a weight on top (page 65) and leave in a cool place for 24 hours. Then store in the refrigerator until required for use. To use, peel off the foil carefully and turn the pâté out (page 65).

Chicken Liver Terrine

550 g ($1\frac{1}{4}$ lb.) chicken or mixed poultry livers
125 ml ($\frac{1}{4}$ pint) port, Madeira or sweet sherry
350 g (14 oz.) pork sausage meat
4 rashers back bacon without rinds
3 slices ham
75 ml (3 fl. oz.) white wine
Grinding of black pepper
Salt to taste
500 g (1 lb.) pork fat, sliced

Marinate the livers in the port, Madeira or sherry for at least two hours. Mash the sausage meat. Chop the bacon and ham very finely and add them to the sausage meat. Pour in the port, Madeira or sherry and the white wine and mix thoroughly. Lastly, chop half of the livers and add them to the mixture. Add seasoning to taste.

Line a terrine with pork fat, keeping aside enough to cover the dish. Put in an even layer of the sausage meat mixture. Cover with a layer of livers. Repeat these layers until the terrine is full, ending with a layer of sausage meat mixture. Cover this with another layer of pork fat. Seal the terrine securely with foil and a lid. Place it in a pan of very hot water and bake at 170°C, 325°F, Gas 3 for about $1\frac{1}{4}$ hours. Remove from the oven and from the pan of water. Place a weight on top which fits the dish as described on page 65 and leave in a cool place for at least 24 hours. Then store in the refrigerator until required for use. To use, peel off the foil and the top pork fat carefully. Serve in the dish, removing the pork fat which lines the terrine when slicing the meat.

Useful Reading

Domestic Preservation of Meat and Poultry
   HMSO. SO Code 24-262. CBH 31094. Prepared by the Ministry of Agriculture, Fisheries and Food and the CIO

Home Curing of Bacon and Hams
   Bulletin 127 of the Ministry of Agriculture and Fisheries in conjunction with the Small Pigkeepers' Council. HMSO 1943

Home Curing
   Bulletin of the British Bacon Curers' Federation

The Home Book of Smoke Cooking
   Sleight and Hull. David and Charles 1973

The Complete Home Freezer
   Norwak. Ward Lock 1975

Meat for your Freezer
   Richards. Faber 1974

Farmhouse Fare
   Countrywise Books Ltd 1966

Charcuterie and French Pork Cookery
   Grigson, Penguin Books 1970

Consumer Guidance Organizations

British Bacon Curers' Federation
Icknield Way
Tring
Herts. HP23 4JY

College for the Distributive Trades
   Food Commodities, Meat Division
Eagle Court
Smithfield
London EC1

Long Ashton Research Station
Department of Agriculture and Horticulture
University of Bristol
Bristol BS19 9AF

Meat and Livestock Commission,
   British Meat Service
PO Box 44
Queensway House
Bletchley
Milton Keynes MK2 2EF

National Federation of Women's Institutes
39 Eccleston Square
London SW1